Renata Azevedo Xavier
Olímpio R. Cardoso

# Ecohydrodynamic Modeling as an Aid to Wetland Management

Renata Azevedo Xavier
Olímpio R. Cardoso

# Ecohydrodynamic Modeling as an Aid to Wetland Management

## Development of Habitat Suitability Indices for Banhado do Taim (ESEC TAIM-RS)

ScienciaScripts

**Imprint**
Any brand names and product names mentioned in this book are subject to trademark, brand or patent protection and are trademarks or registered trademarks of their respective holders. The use of brand names, product names, common names, trade names, product descriptions etc. even without a particular marking in this work is in no way to be construed to mean that such names may be regarded as unrestricted in respect of trademark and brand protection legislation and could thus be used by anyone.

Cover image: www.ingimage.com

This book is a translation from the original published under ISBN 978-613-9-65381-2.

Publisher:
Sciencia Scripts
is a trademark of
Dodo Books Indian Ocean Ltd. and OmniScriptum S.R.L publishing group

120 High Road, East Finchley, London, N2 9ED, United Kingdom
Str. Armeneasca 28/1, office 1, Chisinau MD-2012, Republic of Moldova, Europe
Printed at: see last page
**ISBN: 978-620-7-77410-4**

# SUMMARY

To my parents, biologists at heart.

# ACKNOWLEDGMENTS

To my advisor Rutinéia Tassi for the opportunity to get to know and conduct research at ESEC Taim, and for the challenge of working with modeling.

To CAPES, for the scholarship.

To all the staff at ESEC Taim and ICMBio for their total support for the research and for caring about us when we were in the field.

To my colleagues Vanessa Camargo and Nàjila Rocha for their company and for joining me on the expeditions to Banhado do Taim. Companions for traveling more than 8 hours by bus, walking in the rain, in the sun, snacking on the side of the road, getting mosquito bites (many bites, in fact). You were fundamental... Thanks girls!

To all my colleagues in the 1st[a] master's class in Environmental Engineering at UFSM, a more than multidisciplinary group, especially my colleagues Rafael Cardoso, Daniela Rodrigues, Diego Balestrin, Elis Deon, Graciela Disconzi and the warrior Alice Falleiro, friends on the journey.

To all my colleagues at DESA, in the GERHI, GHIDROS and Ecotechnologies research groups, for the exchange of ideas and experiences, help and partnerships.

To all the friends who provided moments of relaxation and helped hold me up in difficult times.

To my fiancé, work colleague, fellow master's student, friend, companion... Olimpio Rafael Cardoso for sharing all these moments with me, for his encouragement, affection and understanding.

To my family, especially my parents and my brother for their total support and affection, and for your understanding of the times when I was absent.

To God for strength and courage, comfort and hope.

**Thank you very much!**

"When human beings understand that they are part of the world, many transformations will take place. They will no longer throw garbage into lakes, rivers and seas because they will know that these are the very waters that flow through their bodies."

(T. P. Zylberberg)

## SUMMARY

Master's thesis

Postgraduate Program in Environmental Engineering
Federal University of Santa Maria

## ECOHYDRODYNAMIC MODELING AS A SUBSIDY FOR WETLANDS MANAGEMENT

Author: Renata Azevedo Xavier
Supervisor: Rutinéia Tassi
Date and Place of Defense: Santa Maria, January 12, 2015.

Banhado do Taim, located in the Taim Ecological Station (ESEC), is an important wetland in southern Brazil, and is fundamental to maintaining the ecological stability of the region. The main economic activity around the ESEC is the practice of planting irrigated rice through flooding, a process that demands large amounts of water, which are captured from the region's lagoons in the summer months, causing the levels of the lagoons and marshes to drop, which can be detrimental to species dependent on this wetland environment. One of the ways of investigating how species are impacted by this activity is by integrating information on the environmental conditions necessary for the survival of species and their interdependence, through the method of habitat suitability indices (HSAs). This method is considered a form of ecological niche modeling, which can help to improve or develop strategies for managing water resources and conserving biodiversity, and has been widely used in wetland regions. Thus, using this methodology, this work was developed with the aim of adding information to the research carried out by Tassi for Banhado do Taim in 2008, through the devclopment and validation of HIAs, which relate hydrological and habitat aspects, aimed at proposals for managing the site's water resources. To this end, five new AIs were developed for the selected indicator species: reed bunting (*Scirpus californicus*), black-capped sandpiper (*Plegadis chihi) and* loggerhead (*Mycteria americana)* birds, and the reptiles yellow-bellied *caiman* (*Caiman latirostris*) and tiger turtle (*Trachemys dorbigni*). The species designated as wetland indicators were those that could be observed in the field, and hydrological information was available. The IAHs of the species were constructed using information on the height of the water table needed to guarantee the conservation of the species and the area needed to be kept free from flooding during the reproductive period, based on the premise that if ideal conditions were maintained, these would ensure their chemical processes, food availability and reproductive success. Using an environmental quality spatialization model developed especially for the Banhado do Taim, the information from the IAHs was reproduced using thematic maps with a color scale, which reflect the environmental suitability for the species under different scenarios. By analyzing the AIs and the permanence curves, the ideal minimum and maximum water levels for the indicator species of Banhado do Taim were obtained, corresponding to 1.75 m and 3.10 m, respectively. Water levels in this range would guarantee a favorable environment to ensure the reproduction of species and a favorable environment for warm-season migratory species. This, on the other hand, would limit the withdrawal of water for irrigation in conditions of lower water availability, but the use of this resource could be increased in situations that exceed the limit of the maximum level established for Banhado do Taim.

**Keywords**: Ecological Niche Modeling. Ecohydrology. Wetlands. Sustainable management of natural resources.

# 1. INTRODUCTION

Water availability and quality are of fundamental importance for the survival of living beings, as well as for the proper functioning of ecosystems. However, the integrity of the world's water resources is under increasing threat, due to the increase in human population, the expansion of industrial and agricultural activities, in a scenario where climate change could cause major alterations in the hydrological cycle (ANA, 2011).

Human interference in freshwater ecosystems has modified the natural physical and biological characteristics of these environments, altering their productivity and resilience (NILSSON *et al.*, 2007). In this sense, ecohydrology seeks to integrate the knowledge of hydrology and ecology, creating a broader approach to understanding the dynamics of water resources, facilitating the planning and actions of management programs (ANA, 2011).

According to Marengo and Silva (2006), there are major limitations in scientific knowledge regarding the establishment of functional relationships between hydrology, soil, climate and biodiversity in Brazil's main ecosystems. Integrating this information is essential for modeling the spatial distribution of ecological communities, determining the physical structure of the environment and its biota, quantifying natural resources and providing support for projections of future scenarios. The author also states that in Brazil, the relationship between biodiversity and water regime is still little known in seasonally flooded ecosystems such as wetlands.

There are different concepts for wetlands, so this category ends up encompassing different ecosystems, but with the common characteristic of being permanently or temporarily floodable areas (SAITO, 2006), regulated mainly by the annual cycle of floods and droughts. The distribution and diversity of species in these environments vary according to the different habitats arranged along the flood gradient as a result of the hydroperiod[1] , which fluctuates according to the intensity, seasonality and permanence of the floods (MITSCH; GOSSELINK, 1993).

Wetland environments have long been regarded as economically unproductive areas, and drainage or landfill of these regions has been encouraged (BURGER, 2000) and, as a consequence, more than half of the world's wetlands have been lost. As water level fluctuations and flow patterns play a definitive role in the creation and maintenance of different habitats, any activity that alters the hydroperiod to which the flooded area is subjected can also cause changes to the ecosystem (MOTTA- MARQUES; IRGANG; GIOVANNINI, 1997; ZEDLER; KERCHER, 2005).

One of the ways of guaranteeing fluctuations in water levels and mitigating actions that cause alterations to the hydroperiod is through wetland water resource management processes. The management of a natural, economic or socio-cultural environmental resource consists of the articulation of a set of actions by different agents, with the aim of making the use, control and protection of this environmental resource compatible, disciplining the respective anthropic actions, in accordance with the policy established for it, in order to achieve sustainable development (SILVA; PRUSKI, 2000).

As the process of managing wetlands involves various actors from different spheres, there are many interests, often divergent, that drive them (NETO; PINTO; CAMPOS, 2011). Furthermore, in many cases management policies and practices do not take into account the interconnections and interdependencies between water and ecosystems. The importance of water resources and wetlands needs to be recognized and integrated into decision-making in order to meet future social, economic and environmental needs.

One of the ways of integrating this knowledge is through habitat suitability indices (HSA), which seek to encompass information on the actual or potential distribution of a population in the environment through occurrence data, together with the specific conditions and resources necessary for its survival. These indices help to manage ecosystems and mitigate environmental impacts by providing information on the quality of the habitat of a particular species or group of species, helping to improve strategies for managing water resources and conserving biodiversity.

The Taim Ecological Station (ESEC Taim), the area defined for this study, is recognized worldwide as one of the most important conservation areas, preserving wetlands (especially the Banhado do Taim), lagoons, fields, dunes and forests, which are home to a great diversity of plant and animal species (MMA, 2013). The Taim ESEC is one of the fully protected conservation units, according to Federal Law 9985 (BRASIL, 2000), which aim to preserve the area and carry out scientific research.

The Banhado do Taim (BT) is essential for maintaining the ecological stability of the region. Among the functions attributed to the wetland are food production, biodiversity conservation, flood containment and pollution control. The most important processes in this ecosystem are plant production and the storage of nutrients, water and biodiversity (MMA, 2013).

The main economic activities around the ESEC are irrigated rice farming, forestry and livestock farming. In its surroundings there are small agricultural properties (SCHAFER *et al., 2009).* There is also a history of conflicts in the region involving fishing and hunting, as well as several incidents of animals being run over (SETE, 2010; TASSI, 2008). Rice crops are irrigated by flooding, a process that uses a large amount of water in the summer months, causing a significant reduction in the level of the Mangueira Lagoon (a tributary of the BT) during periods of low rainfall, which ends up affecting the water level inside the BT (MOTTA MARQUES; VILLANUEVA, 2001).

In addition, Banhado do Taim is vulnerable to extreme climatic situations, as reported by Palo Jr:

In the last five years, I have witnessed two atypical situations for the Taim, which have led to the death of many animals and significant changes in the vegetation. A flood in 1984 inundated the entire area of the ecological station and neighboring lands, causing the death of dozens of tachàs, victimized by the cold, due to the loss of thermal insulation, and starvation. In the summer of 88/89, an inexorable drought punished the reserve, depleting almost all the water in the bays and lagoons, which is the very essence of the Taim (PANESSA; PALO JR.; TRESCA, 1989, p. 2).

The management of shared uses at ESEC Taim has been discussed for years. Studies carried out at ESEC Taim (MOTTA-MARQUES *et al.*, 2002; PAZ, 2003; TUCCI, 1996; VILLANUEVA, 1997) have helped to provide some guidelines for managing water resources, based on the realization that actions outside the wetland directly affect the quantity of water and the preservation of its environmental conditions (VILLANUEVA *et al.*, 1997). However, many of these studies essentially consider hydrological criteria.

Motta-Marques, Irgang and Giovannini (1997) advocate that the identification of the hydroperiod and its relationship with the biotic components and functions of the ecosystem are the basis for studies linked to the management of these ecosystems, thus guaranteeing their stability in order to preserve their functions.

The aforementioned authors considered it essential that there should be a variation in the water levels of the BT ecosystem, reproducing the hydroperiod that is the signature of the wetland, as maintaining permanent levels could harm the driving force of the wetland. It is also advisable to establish a limit level, below which there should be no exploitation, thus ensuring the conservation of this ecosystem by allowing the existence of a hydroperiod closer to the natural one and suited to the requirements of the species dependent on the habitats within it.

In 2008 , a new study sought to incorporate eco-hydrological information into the BT's water resource management rules, through the use of AHI (TASSI 2008). This study identified the need for a better understanding of the relationship between species and the hydrological regime of the ESEC Taim and, therefore, AHI were developed for some biological indicators.

This study sought to add information to the work of Tassi (2008) by developing new IAHs for the BT and using them to revise water resource management rules. The aim is to ensure the multiple use of water together with the preservation of the ecosystem and biodiversity, with a view to its stability and guaranteeing the maintenance of its ecological, economic and cultural functions, while also collaborating with the management plan for the Conservation Unit.

## 2. BACKGROUND

Habitat degradation is considered an unintended consequence of water resource management, resulting from a lack of understanding of the water flows needed to sustain freshwater ecosystems. The prevention and mitigation of environmental degradation, as well as the rehabilitation of degraded environments, have been increasingly incorporated into environmental management plans to guarantee the integrity of these areas (RICHTER *et al.*, 2003).

Ensuring the quality, conservation and protection of water resources through appropriate management measures and instruments are essential to support all aspects of human life and the associated terrestrial and aquatic ecosystems. According to Federal Law 9433 of January 8, 1997 (the Water Law), the management of water resources must provide for the multiple use of water, and its use must be rational and integrated, always aiming for sustainable development.

Despite the importance of the goods and services provided by wetland ecosystems, their global status is worrying. It is estimated that during the last century more than half of the world's wetlands have been lost as a result of human activities, through drainage, urbanization and agriculture (IUCN, 2000). The continued degradation of wetlands can lead to the loss of biodiversity, alterations in ecological functions and environmental services, as well as affecting human health, community livelihoods and economic activities (BARBIER; ACREMAN; KNOWLER, 1997).

According to Bobbink (2006), in some densely populated regions of Europe, North America and East Asia, more than 80% of wetlands have been lost or severely degraded. In Brazil, the wetlands of the South and Southeast (which concentrate 60% of the country's population) are facing a growing threat of water scarcity as a result of economic growth and accelerated urbanization. These considerations exemplify how threatened these ecosystems are, emphasizing the need to improve water quantity and quality management in order to guarantee the preservation and conservation of these environments.

Social and economic factors have been primarily responsible for environmental changes in wetlands, through the attempt to maximize benefits by overexploiting resources. Many of these areas have been extensively modified to increase their agricultural productivity.

Agricultural activity, both at the local level and at the watershed level, has significant effects on ecology and therefore on the functioning of ecosystems and the modification of hydrological flows. Other factors that exert negative pressure on wetlands are the construction of dams on rivers, excessive water withdrawal, groundwater abstraction, pollution through the use of fertilizers and pesticides (MCCARTNEY *et al.,* 2010).

Specifically in the southern region of Brazil, wetlands, or banados, are common formations of the pampa biome. These ecosystems used to occupy large stretches of the coastal zone and also more inland regions of Rio Grande do Sul (BURGER, 2000). However, they can currently be considered vulnerable and threatened environments, due to urban growth, silting, drainage and pollution (CARVALHO; OZORIO, 2007), with emphasis on the rice farming activities that take place especially in regions that are characterized as wetlands, or close to them.

The results of rice farming on wetland systems are the reduction of natural ecosystems, soil compaction, porosity reduction and salinization, variations in the water table, eutrophication, erosion and silting up of water resources, contamination by pesticides, risks arising from monoculture, shortage of water supply, excessive use of energy, excessive use of water, as well as reducing the potential for other uses of the area, such as tourism, leisure and fishing (CHOMENKO, 2007).

Ecosystem management must ensure the protection of populations of all species, guarantee the maintenance of their functions and recognize that human beings are part of the environment (PRIMACK; RODRIGUES, 2001). One of the ways of assessing these environmental connections is through tho use of bioindicator organisms, which help to diagnose the causes of environmental impacts, determine the condition of an environment in the face of disturbances, and assess the effectiveness of mitigating actions (DALE; BEYELER, 2001).

It is therefore necessary to implement measures to guarantee the conservation and protection of wetland ecosystems by creating appropriate policies for the sustainable management of water resources, in order to support all aspects of human life and the associated terrestrial and aquatic ecosystems. In this context, one of the methods that has contributed to environmental conservation and management processes is ecological niche modeling, which in many cases also uses indicator species.

# 3. OBJECTIVES

To develop new Habitat Suitability Indices for the Banhado do Taim, in order to complement the previous study developed by Tassi (2008). Also, to generate information that will help to draw up proposals for managing the region's water resources, with the recommendation of a minimum and maximum water level to be maintained in order to guarantee the conservation of the indicator species, as well as the wetland ecosystem.

## 4. WETLANDS, CONCEPTS AND ENVIRONMENTAL SERVICES

Wetland ecosystems are considered a natural resource of global importance, mainly due to the great diversity of plants and animals, especially birds (BOBBINK *et al.*, 2006). For this reason, the protection of these environments has become a priority worldwide, supported by international agreements such as the Ramsar Convention (the result of the Ramsar Treaty, a global intergovernmental agreement aimed at conserving wetlands) and the International Convention on Biological Diversity.

There are several concepts for the term "wetlands", which ends up bringing together a series of distinct ecosystems, however, the common characteristic of these ecosystems is temporary or permanent flooding, through water from rainfall, the water table, surface runoff and contributions from rivers and lakes. These are areas where water is the main factor regulating the environment and the biota (ZEDLER; KERCHER 2005), and its accumulation on the surface depends on the type of relief, water availability and the type of soil (SAITO, 2006).

The most accepted concept is that established by the Ramsar Treaty (1971), which establishes as wetlands the entire extent of marshes, ponds and peatlands, or surfaces covered by water, whether natural or artificial, permanent or temporary, containing standing or flowing water, fresh, brackish or salt, including marine areas with a depth of up to six meters at low tide (PRATES *et al.*, 2010).

FEPAM (Fundaçâo Estadual de Proteçâo Ambiental Henrique Roessler), the institution responsible for environmental licensing in Rio Grande do Sul, uses Junk's (1993) definition of wetlands. According to this author, wetlands are terrestrial-aquatic transition zones that are periodically flooded by the lateral reflection of rivers and lakes and/or by direct precipitation or groundwater and which result in a particular physical-chemical environment that leads the biota to respond with morphological, anatomical, physiological, phenological and/or ethological adaptations and to produce community structures that are characteristic of these systems. Regionally, wetlands are also called "banados", which comes from the Spanish "banados" (BURGER, 2000).

Among the goods and services provided by these ecosystems to society and the environment are the storage and cleaning of water, recharging the water table, regulating the local climate, maintaining biodiversity, regulating biogeochemical cycles, storing carbon, protecting coastal areas from rising sea levels and controlling erosion, as well as providing different habitats for numerous species. These environments also provide various subsidies for traditional human populations, through fishing, subsistence farming, and extensive livestock farming, have tourist value and great relevance for leisure and recreation (BOBBINK *et al.*, 2006; CHOMENKO, 2007; JUNK *et al.*, 2011).

The environmental services provided by these ecosystems also sustain and maintain the planet's essential life processes, such as primary production and nutrient circulation. Each of these sustaining services is essential for human well-being, whether the services are considered at a local, regional or global level (SCDV, 2006). Therefore, the rational use of wetlands consists of their sustainable use for the benefit of humanity, in a manner compatible with the maintenance of the natural properties of the ecosystem, in such a way as to produce the greatest ongoing benefit for current generations, while maintaining their potential to satisfy the needs and aspirations of future generations (RAMSAR, 2002).

## 4.1 Wetland Hydrology

The hydrodynamic processes of wetland ecosystems are mainly regulated by the cycle of floods and droughts. The processes of species distribution and diversity vary between the different habitats arranged along the flood gradient, due to the intensity, seasonality and permanence of the floods, criteria that characterize the hydroperiod of a wetland (FERNANDES *et al.*, 2010).

The hydroperiod is determined by the periodic or regular occurrence of flooding or saturated soil conditions (frequency, duration), as a result of the water balance, topography and subsurface conditions. The functions of a wetland can be modified by changes in the hydroperiod, also causing changes in the pattern of communities, productivity and habitats (MOTTA-MARQUES, 2002).

For these functions to be maintained, it is necessary to identify and define the specific hydroperiod or series of hydroperiods, so that this information can be used to manage the ecosystem (MOTTA- MARQUES, 1997). Changes in the hydroperiod of a wetland can be attributed to natural events, seasonality, or anthropogenic actions, such as the withdrawal of water through drainage and irrigation (MOTTA-MARQUES, 2002).

Some wetlands are subject to cycles of 5 to 20 years of drought. In these cycles, the wetland is reduced and large open areas are exposed, followed by seed germination in substrate banks exposed to re-flooding after rains, with the development of perennial species (MITSCH; GOSSELINK, 1993). This germination phase is only possible because wetland substrates act as ecosystem memories and can then restore basic structures and functions when the necessary conditions are created (MOTTA-MARQUES, 1997).

In cases of extreme drought and flood conditions, if these are maintained for prolonged periods, they can cause the extinction of the wetland habitat function, which requires oscillating levels between a maximum water level and a minimum water level, at frequencies and durations appropriate to the demands and tolerance of the biota (MOTTA-MARQUES, 1997).

The structure and functioning of wetlands with fluctuating water levels is described by the concept of the flood pulse (JUNK; WANTZEN, 2004), which occurs as a result of the processes of flooding and drought. This concept defines, analyzes and explains the lateral exchange of water, nutrients and organisms between rivers or lakes and their connected wetlands, defining processes and patterns of hydrological signatures in the change of ecological conditions between the terrestrial and aquatic phases. Also included in this concept are areas constantly flooded by rainfall and the periodic rise of the water table.

Any activity that alters the hydroperiod to which the wetland is subjected can cause changes to the ecosystem, resulting in natural disturbances and increased environmental degradation (ZEDLER; KERCHER, 2005; ZACHARIAS; DIMITRIOU, 2010).

## 4.2 Biodiversity of Wetlands

The term biodiversity refers to variation between organisms and ecological systems at all levels, including genetic variation in populations, morphological and functional differences between species and variation in biome structure and ecosystem processes in both terrestrial and aquatic systems (RICKLEFS, 2010).

Wetlands have high productivity and biodiversity and are among the most productive ecosystems on the planet, being compared to tropical forests (RICKLEFS, 2010). They are characterized by a wide variety of physical and chemical conditions, spatial heterogeneity and high nutrient availability (BARBIER; ACREMAN; KNOWLER, 1997).

Gopal and Junk (2000) define wetland species as all plants, animals and micro-organisms that inhabit permanently or periodically flooded areas (including migratory individuals from adjacent or distant habitats), or that depend directly or indirectly on the wetland habitat or any species present in this environment.

The water level and the properties of the runoff influence the development of vegetation, which interferes with local hydrology, soils, the production of organic matter, as well as forming the basis for all other forms of life, making them essential for the performance of all wetland functions (PAZ, 2003).

Most wetland plant communities are made up of aquatic macrophytes, which vary according to the hydrological regime, morphometry and other physical characteristics of each system (SCHWARZBOLD; SCHAFER, 1984). These plants have morphological adaptations for flooded environments, and there is also marshy herbaceous vegetation, ranging from aquatic species to amphibians, whose occurrence will depend on the condition of greater or lesser flooding in depth and duration (JUNK, 2013).

According to Cordazzo and Seeliger (1988), aquatic macrophyte communities play an important role in substrate stability, preventing erosion, representing habitat for various organisms and constituting an important source of detritus for estuarine food webs. In order to survive in this hostile environment, plants have developed anatomical, physiological and biochemical adaptations, and are also important as a refuge and food source for various species of invertebrates and vertebrates.

Aquatic macrophyte populations are profoundly affected by seasonal environmental changes, especially rainfall and fluctuations in water levels (ESTEVES, 2011). Studies of macrophyte communities in flooded areas have found that unnatural or extreme changes in water levels can cause physiological changes, lead to the loss of aquatic macrophyte species and interfere with the abundance of individuals (ZOHARY; OSTROVSKY, 2011).

Various organisms depend on these flooded environments for part or all of their life cycles, for obtaining food, water and shelter, and for migration and reproduction processes. The combination of surface water accumulation, high nutrient levels, organic matter degradation and high rates of primary productivity are ideal for the development of a complex food chain. This favors producer organisms such as phytoplankton and plants, as well as organisms that form the base of the food chain, such as zooplankton, aquatic invertebrates and small fish. Many of these invertebrates and fish serve as food for larger predators, such as amphibians, reptiles, fish, birds and mammals (WETLANDS INTERNATIONAL, 2009).

Borkhataria *et al.* (2012) point out that human-caused changes in the temporal patterns of wetland hydrology can have important consequences for the migratory process and demography of wetland birds. As a way of ensuring the conservation of waterfowl, it is important to manage water levels in wetlands (BORKHATARIA *et al.*, 2012; U.S. Fish and Wildlife Service, 1999).

Wetlands also have interrelationships with nearby environments through ecological processes such as the migration of fauna, the dispersal of vegetation and the exchange of sediments carried by the wind and water flows (IBAMA, 2000). However, they have their own characteristics depending on the chemical and physical conditions present, the type of vegetation and the existing fauna (CARVALHO; OZORIO, 2007).

## 5. ECOLOGICAL NICHE MODELING

The concept of niche has evolved over time, and Hutchinson (1957) formalized it by defining it as "the n-dimensional hypervolume limited by interactions with other organisms, which involves all physiological responses to environmental conditions, and depends on the availability of resources, under which populations show a positive growth rate".

Hutchinson (1957) further subdivided this classification into the fundamental niche (physiological or potential) - which refers to all the environmental conditions that allow the species to grow and reproduce - and the realized niche (ecological, actual) - where the effects of competition reduce the fundamental niche of a species, or the area it can occupy.

While the ecological niche relates a set of environmental variables to the suitability of species for the environment, the processes of ecological niche modeling, habitat suitability modeling (also called predictive species distribution modeling (LOBO; JIMÉNEZ-VALVERDE; REAL, 2008; GIANNINI *et al.*, 2012) relate environmental variables to the probability of occurrence of species (HIRZEL; LAY, 2008). This methodology has focused on issues related to the niche of a given species, the selection of environmental variables that act on the species, response curves to the quality of the environment, and the interactions between the variables (HIRZEL; LAY, 2008).

Grinnell (1917) listed several factors that potentially affect the distribution of species in an environment. These include local vegetation, food availability, reproduction, places of refuge, interspecific effects, climate, soil, topography and individual preferences. The researcher also points out that species do not react uniformly in an environment, and what defines their distribution is always a combination of factors.

In order to understand the geographical distribution and abundance of a species, we need to know its evolutionary history, the resources necessary for its survival, birth, mortality and migration rates, intra- and inter-specific relationships and the effects of environmental conditions. A species can potentially occur and persist in a given location as long as there are certain conditions within acceptable limits and resources necessary for the species (RICKLEFS, 2010).

The presence and success of an organism or group of organisms depends on this set of conditions, and any factor that approaches or exceeds the tolerance limits becomes a limiting factor (PRIMACK; RODRIGUES, 2001). Examples of limiting factors are temperature, humidity, pH, salinity and the concentration of pollutants and water (PERONI; HERNANDEZ, 2011). Water is considered a determining factor for terrestrial or aquatic environments where its quantity is subject to great fluctuations (ODUM, 2004), as in the case of wetlands.

Ecological niche modeling takes place by integrating information on so-called indicator species, which refers to organisms with specific characteristics, such as sensitivity to small environmental changes, small tolerance limits and easy identification (ODUM, 2004), and the factors that condition them through mathematical and statistical methods.

This information can be incorporated into Geographic Information Systems (GISs) by means of algorithms (finite sequences of commands, executed through code written in a programming language) specific to predictive distribution modeling (GIANNINI *et al*,

2012). As a result, thematic maps are obtained which have a color scale that indicates a greater or lesser degree of environmental suitability for a given species to be present in a given area, depending on the variations of the defined limiting parameter (SIQUEIRA, 2005).

Ecological niche modeling processes have also encouraged the creation of various websites with the aim of integrating information obtained by researchers and institutions. These sites have regional, national and global databases of species occurrences, linked to scientific collections, as well as sites with information on variables directly linked to species distribution, such as temperature, rainfall and topography.

The models obtained for each indicator species (based on current conditions) should be tested with field data, considering the most different situations (O'NEIL *et al.*, 1988). These data can then be re-executed under a variety of predictions of future scenarios of changes in the environment, in order to estimate the ecological vulnerability of the site (LOPEZ; HEGGEM, 2001).

Ecological niche models have helped in conservation and environmental management processes, making it possible to delimit the ecological requirements of species and their limiting factors, understanding biogeographic barriers, identifying sites for the reintroduction of fauna, designing conservation plans, predicting the effects of habitat loss and the effects of climate change (PETERSON, 2006).

One form of ecological niche modeling that has helped in the development of natural resource management processes, especially with regard to wetland water resources, is the habitat suitability index method, which is the subject of the following section.

# 6. HABITAT SUITABILITY INDEX (IAH)

The Habitat Suitability Index, an ecological niche modeling method, was developed in the 1980s by the U.S. Fish and Wildlife Service (U.S. Fish and Wildlife Service, 1980), and is based on the functional relationships between the biota and the abiotic variables of an ecosystem, assuming that the quality and quantity of a habitat can be described numerically, allowing impacts on species to be assessed.

This institution has drawn up a series of IAHs with the aim of restoring the Everglades region in Florida, an important wetland region in the south of the United States, threatened by environmental degradation, which has resulted in the loss of half of its natural area, damage to goods and services, as well as the local extinction of species (CISRERP, 2014).

To this end, HIAs were developed for around one hundred species indicative of this wetland environment, and the information obtained was used to improve habitat assessment processes and quantify the effects of ecosystem management (ROLOFF; KERNOHAN, 1999). The environmental variables considered to be limiting factors were local hydrodynamics, availability of food resources, vegetation cover and the conditions necessary for species reproduction.

The creation of HAIs is related to the occurrence, permanence and distribution of populations in the environment, and aims to portray the specific conditions and resources needed by a particular species to ensure its survival and reproduction. As the aim of using HAIs is to maintain the stability of an environment and conserve its functions, it is important to use indicator organisms that are representative of the ecosystem under study, since an impact on bioindicator populations can also affect the entire community.

To develop the AHI, the following parameters can be used as eco-hydrological conditions: water quantity and quality, fluctuations in water levels, magnitude of flooding, frequency and duration of flooding, local topography, distribution and vegetation cover (LUZ, 2002; TARBOTON *et al.*, 2004; MORA; MAGER and SPIELES, 2011), among others.

Once the indicator species and the environmental variables that affect these species have been selected, functions are established that vary in a range between 0 (unfavorable environment for the indicator) and 1 (totally favorable environment for the indicator), depending on the variable adopted (BROWN; HARTWICK, 1988; TARBOTON *et al.*, 2004), as exemplified in figure 1, for a wading bird, and in figure 2 for a hypothetical species of fish.

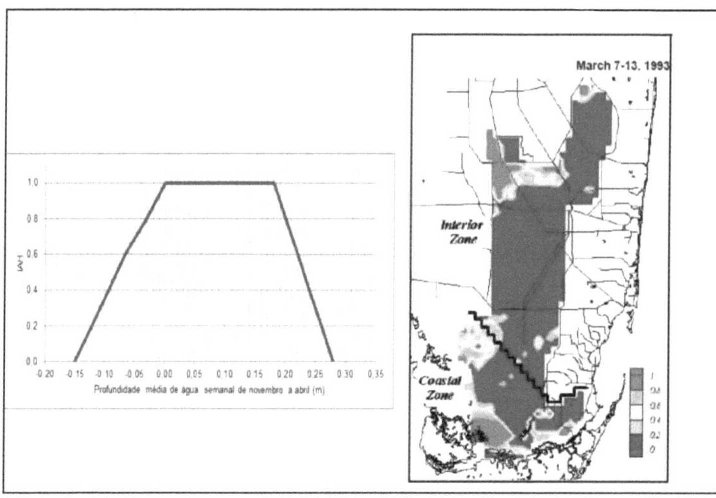


Figure 1 - Habitat suitability for Everglades wading birds, as a function of average weekly water depth for the months of November to April. Figure 1a - Suitability index. Figure 1b - Habitat suitability map generated from the HAI. Adapted from Tarboton *et al.*, 2004.

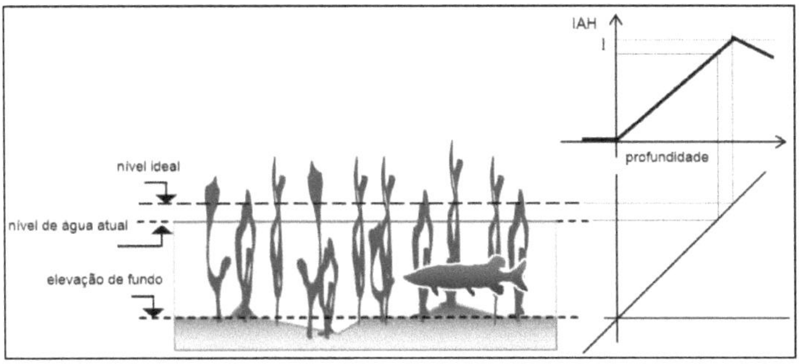


Figure 2 - Schematic representing the relationship between the environmental variable water depth and the AHI of a hypothetical species of fish. Adapted from Luz (2002).

In some cases, it may be necessary for the same species to establish more than one AHI function for different variables, such as food availability, distance from a body of water, depth and percentage of non-flooded area. These variables can be combined to form a single value (TARBOTON *et al.*, 2004). Table 1 seeks to elucidate the steps to elaborate the composite AHI for the turtle species *Trachemys scripta*, according to the point of view of the researchers Morreale and Gibbons (1986). Among the variables that should be assessed in the construction of this AHI are the relationship between the physical variables of the habitat, the resources used by the species, such as food and water, and different habitats in terms of their vegetation cover.

Table 1 - List of habitat variables, resources and type of cover required to calculate an IAH value for the turtle species *Trachemys scripta*

| Habitat variables | Requirements | Type of cover | |
|---|---|---|---|
| Percentage cover of emergent and submerged vegetation | Food/ Coverage | - Wetland with deciduous forest;<br>- Tree/shrub wetland;<br>- Herbaceous wetland;<br>- Riparian vegetation;<br>- Lake vegetation. | IAH |
| Speed | Water | | |
| Water depth | | | |
| Water regime | | | |
| Water temperature | Temperature | | |

Source: Adapted from Morreale and Gibbons (1986).

The use of more than one HAI to assess habitat quality is an attempt to reproduce more faithfully the environmental conditions that are the result of different local attributes. However, developing composite indices is a difficult process due to the need for extensive knowledge of the species and its dependence on the complex phenomena that occur between the variables (TASSI, 2008).

As an example of the application of AIs in wetland management processes, we can cite the study carried out by Mushet, Euliss and Stockwell (2012), in the state of North Dakota, United States, where AIs were developed for five amphibian indicator species, depending on variables such as percentage of available wetland, percentage of soil type and distance from farmland. By analyzing these AIs, the authors found that changes in the environment have a strong influence on indicator species. It was therefore found that it is essential to conserve these areas, imposing limits on land use for urbanization or agricultural practices.

Another study, carried out by Babbitt (2005) in the United States, in the New Hampshire region, related the richness of anuran species in the construction of IAHs to the size of the wetland and its hydroperiod. The author came to the conclusion that these factors are all interconnected, emphasizing that the composition of the anurofauna changes according to the hydroperiod. As some wetland management policies use the size of the area as a parameter to protect it, the author proposes that even small areas should be protected to guarantee the maintenance of local biodiversity.

In the same year, 2005, a similar study was carried out in New England (SKIDDS; GOLET, 2005), relating the richness of the anurofauna and the suitability of the wetland hydroperiod, in order to assess the minimum period of flooding necessary to complete the reproductive cycle of each species. The variables used to construct the AIs that showed the greatest influence on the hydroperiod of the study areas were vegetation cover and water depth. The results indicated that it is possible to estimate the hydroperiod of seasonal wetlands and assess their suitability for each species, without the need for a long period of hydrological monitoring. The authors concluded that the method developed can be used by managers to plan and maintain these ecosystems.

Lo Galbo *et al.* (2013), in a case study carried out in the Everglades (Florida, United States), used the floating aquatic macrophyte *Nymphaea odorata* as an indicator species, and developed its AHI as a function of hydrological variables such as depth, seasonality of water levels and precipitation. The authors evaluated the effects of restoring the environment by creating situations through modeling. In order to assess the suitability of this environment, models were created to simulate existing conditions, conditions prior to drainage in the wetland and future conditions, taking into account the restoration of the environment.

The results indicate that the restoration of the Everglades will bring greater benefits to the native vegetation and local biota than the current situation provides, and that due to the anthropogenic action suffered in the region, there have been major changes in the dynamics of the environment, as well as in its biodiversity (LO GALBO *et al.*, 2013).

In Brazil, the use of indicators such as macrophytes, birds and aquatic mammals can be cited for the composition of IAHs, used to define acceptable limits for fluctuations in the Banhado do Taim hydroperiod, allowing the creation of proposals for the sustainable use of water resources (TASSI, 2008), since the site has a history of conflicts over water resources, including conservation of the ecosystem, water withdrawal for rice cultivation, livestock farming, fishing and hunting (NEMA, 2002).

In addition to wetland ecosystems, HIAs have been developed for a wide range of biomes in various countries, with many studies seeking to assess the impacts of anthropogenic influence on these environments. The majority of these studies use vertebrate animals such as fish, amphibians, reptiles, birds and mammals as indicators, as well as invertebrates and native vegetation (RALEIGH; ZUCKERMAN and NELSON, 1986; GUTZWILLER; ANDERSON, 1987; LARSSON, 2003; ARMITAGE; FONG, 2004; BROTONS *et al*, 2004; DUBERSTEIN *et al.*, 2008; BRATTEBO *et al.*, 2009; DONAVAN *et al.,* 2011).

The studies presented show that habitat suitability indices are an important tool for decision-making and have been widely used in wetland regions, as they generate significant information about these ecosystems, helping to revise and improve strategies for managing water resources and conserving biodiversity.

# 7. CHARACTERIZATION OF THE STUDY AREA

In order to draw up HIAs for Banhado do Taim, it is essential to understand the dynamics of the environment, from its geological formation, hydrology, physical phenomena, biodiversity, the pressures and threats on this ecosystem, and its importance for local communities.

All these variables are intrinsically linked, and through a bibliographic review we hope to gain a more specific understanding of the functioning and particularities of this complex ecosystem, one of the few in the state of Rio Grande do Sul that is legally protected as a conservation unit.

## 7.1 Location of ESEC Taim

The Taim Ecological Station, created in 1986, has an area of approximately 11,000 hectares, with a proposal to expand the unit to 33,000 hectares, covering the whole of Banhado do Taim (MMA, 2013). It is located in the southeast of the state of Rio Grande do Sul, on the coastal plain of the municipalities of Rio Grande and Santa Vitòria do Palmar, in the extreme south of RS, in a narrow strip between Lagoa Mirim and the Atlantic Ocean (32° 20' and 33° 00' South and 52° 20' and 52° 45' West) (Figure 3). The ESEC is part of the Mirim-Sâo Gonçalo Hydrographic Basin, which is part of the South Atlantic hydrographic region.

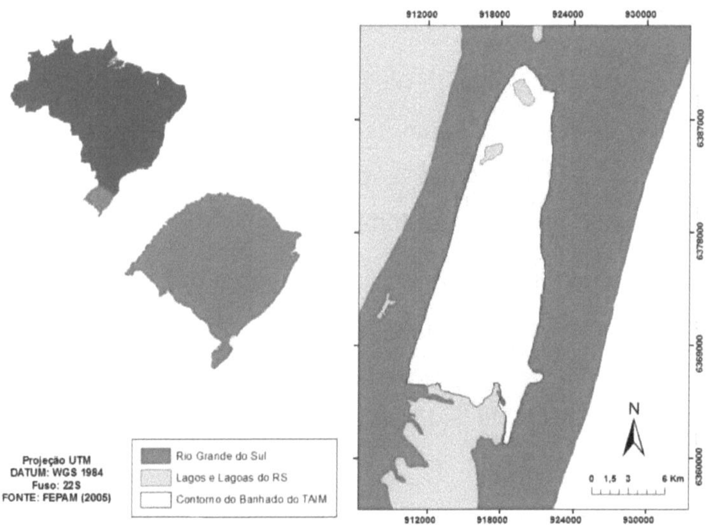


Figure 3 - Location of the Taim Ecological Station, south of the coastal plain of Rio Grande do Sul.

The area of ESEC Taim is characterized by flat terrain with areas that are frequently flooded and others occupied by permanent freshwater marshes and lagoons, with Lagoa Mangueira being the largest. Water levels and flooded areas vary both seasonally and annually. Morphodynamics and the hydrological regime are the most important environmental constraints (BAGER, 1997; TASSI, 2008).

The action of the winds is predominantly in a northeasterly direction, occurring throughout the year, but more active in the spring and summer months, with the west-southwest wind being more active in the winter months. In general, most of the lagoons on the coast of RS are shallow with small relative depths, so they are less prone to thermal stratification and more susceptible to the action of the winds, which act as an agent in the resuspension of matter and mixing of the water column (NIMER, 1977).

## 7.2 Geology of ESEC Taim

The current configuration of the Rio Grande do Sul lagoon system began around 5000 years ago, after a process of marine transgression during the last glacial event, favoring the formation of a sandy barrier (Barrier IV) (VILLWOCK *et al.*, 1986) which isolated a wide strip of lowland between the continent and the ocean. A complex depositional system was established, consisting of various sedimentation environments, thus favoring the formation of lagoons.

During the last Pleistocene eustatic event, the restinga (Pleistocene sedimentary deposits) broke up and partially eroded, in what is now the Banhado do Taim and where the Nicola and Jacaré Lagoons are today, having served as a spillway to the Mirim Lagoon (SCHWARZBOLD; SCHAFER, 1984). This is exactly where the water connection between the lagoon and marsh system and the latter with the Mirim Lagoon occurs today.

The deposition of material brought by the Patos Lagoon in lines parallel to the coast towards the south during the Holocene caused the growth of these bundles to close the connection between the Mirim Lagoon and the sea through the Taim spillway, forming the Mangueira Lagoon. Lagoa Mirim, in turn, aided by the strong erosive processes of Laguna dos Patos over the Pleistocene terraces, established a drainage connection with Laguna dos Patos, through the channel now known as Canal de Sâo Gonçalo (SCHWARZBOLD; SCHAFER, 1984).

The main soil classes in the ESEC-Taim region are Eutrophic Melanic Gleissolo (soils formed by marine or lake sediments, occupying flat terrain at altitudes of between 3 and 5 m) and Neossolo. These soils are sandy, poorly or very poorly drained, with low permeability (PINTO *et al.,* 1999).

## 7.3 Hydrology of ESEC Taim

Climate is one of the factors that helps to control the rates of weathering and erosion on the continents through rainfall, which conditions the surface runoff of water and the transportation of its debris to the coastline, and the levels of evaporation and rainfall. The coastal plain of Rio Grande do Sul has a mild, super-humid mesothermal climate with no dry season, an average temperature of between 16°C and 20°C and annual rainfall of between 1000 and 1500 mm (NIMER, 1977). In studies on climate change, projections show a tendency for temperatures to rise in practically every month of the year in the region of the Taim hydrological system (BRAVO *et al.*, 2011).

The study region is monitored by five weather stations. With regard to precipitation, the longest series belong to stations located close to urban centers (Pelotas, Rio Grande and Santa Vitória do Palmar) and the shortest series belong to stations located in rural areas, but closer to the BT.

Average monthly rainfall is evenly distributed throughout the year, increasing slightly in the winter period. With regard to evaporation, there is a strong seasonal component, with higher evaporation rates during the warmer periods of the year (Figure 4) (TASSI, 2008).

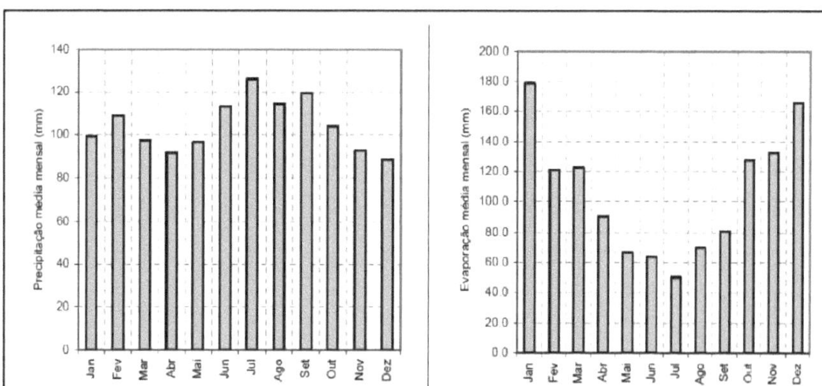


Figure 4 - Seasonality of the rainfall and evaporation series for the ESEC Taim region.

Source: Tassi, 2008.

According to Tucci (1996), the Taim wetland is considered to be a hydrological system essentially governed by levels, under the influence of factors such as precipitation and evapotranspiration, the flow capacity of the channels, the storage capacity of the lagoons and the limitrophic Areas, the levels of the Mirim Lagoon, the flow characteristics of the sets of conduits of the outlet structure and the water withdrawals for irrigation.

The Taim ESEC does not cover the entire hydrographic basin, which influences the water levels of the marsh and the surrounding area. Actions outside the marsh directly affect the quantity of water and the preservation of its environmental conditions. Likewise, any action on the sluice gate system at the outlet of the wetland affects properties outside the ecological station (VILLANUEVA *et al.*, 1997).

The hydrological system of the Taim ESEC has already been the subject of several studies (TUCCI 1996; VILLANUEVA, 1997; MOTTA-MARQUES *et al.*, 2002; PAZ, 2003). The study carried out by Paz (2003) in the ESEC Taim region sought to understand the hydrodynamics of the environment by applying a two-dimensional horizontal hydrodynamic model (IPH-A model), associating a drying/flooding algorithm to the system formed by the Mangueira Lagoon and the wetland.

As a result, it was possible to see that the spatial distribution of the vegetation regulates the general pattern of water circulation in the marsh, and that the factors wind shear action on the water surface, resistance to runoff and protection against wind action provided by the aquatic macrophytes and the interaction with the Mangueira Lagoon, all play a fundamental role in the hydrodynamics of the Taim and are related. Another relevant fact found by Paz (2003) is that, as the water level rises or falls, different regions of the marsh become dry, resulting in different hydrodynamic patterns, which also differ in terms of the influence they suffer from the wind and the Mangueira Lagoon.

The wetland region varies according to the season, where during the rainy season it becomes completely flooded and is practically inaccessible, while in the summer, with less intense rainfall, the flooded portions become more restricted and there is also an expansion of the vegetation along its edges (GOMES; KRAUSE, 1982).

As the distribution of species in the wetland environment and their diversity vary between the different habitats along the flood gradient, variations in water levels can lead to the loss of native species with small tolerance limits to changes in the environment. Changes in the flood gradient can limit connectivity between environments through natural barriers, as well as favoring the process of invasion and the establishment of exotic species (BUNN; ARTHINGTON, 2002).

Information on the bathymetry of Lagoa Mangueira and Banhado do Taim was gathered by the Institute for Hydrological Research (IPH/UFRGS) in March 2003, using relatively sparse topographic data (PAZ, 2003). Although this information was not very precise, it served as the basis for the Numerical Terrain Model (MNT) of the BT, drawn up by Villanueva (1997) and Paz (2003).

In the drainage area of the marsh, outside the season, there are large areas of rice cultivation by flooding. This type of farming uses a large amount of water in the summer months, lowering the lagoons during the period that coincides with the highest evapotranspiration (VILLANUEVA *et al.*, 1997), and consequently affecting the hydrological regime of the BT. According to IBAMA (1993, apud FRAGOSO-JÛNIOR *et al.*, 2007), in the month of January alone, water withdrawal can reach 72 $m^3$ /s for rice crops, which represents approximately a 10 cm drop in the water level of the Mangueira lagoon in 15 days of pumping.

As this water body is a direct tributary of the BT, a variation in level of this order of magnitude can significantly impact species that depend on specific hydrological conditions during this time of year (TASSI, 2008), compromising local biodiversity and ecosystem functions.

## 7.4 Biodiversity of ESEC Taim

In the southern region, wetlands are mainly associated with coastal lagoons, with a wide variety of aquatic macrophyte communities (SCHWARZBOLD; SCHAFER, 1984). The coastal grasslands form a mosaic with the marshes and woodlands and are made up of numerous species of grasses, legumes and cyperaceae, the latter mainly in the flooded areas (BURGUER, 2000) and this is no different in the BT. There are practically no endemic species in this region, despite the great length of the coastal strip, since the flora did not originate through local speciation processes, but through migration from neighboring and geologically older regions (WAECHTER, 1985).

The Taim Ecological Station, located on the south coast of Rio Grande do Sul, is considered to be very heterogeneous in terms of ecosystems, comprising extensive areas of marshland, coastal dunes, oceanic and lacustrine beaches, lagoons (Nicola, Jacaré and part of Mangueira Lagoon), fields and capons of native forest (MAHLER; KINDEL; KINDEL, 1996).

A total of 331 species of aquatic macrophytes have been described for the marshland region, the most representative environment of the Station, consisting of large swampy areas covered in dense and almost uniform vegetation (BURGER, 2000), with the physiognomic highlights being the spadana, *Zizaniopsis bonariensis* (Gramineae) and the marsh rush or tiririca, *Scirpus californicus* and *Scirpus giganteus* (Cyperaceae). At the edge of the marsh, in higher places, the vegetation is shrubby and quite thick, with the cork tree, *Erythrina crista-galli* (Fabaceae), a larger tree, being common (WAECHTER; JARENKOW, 1998). The dynamics and spatial variability of the vegetation in the BT were studied by Ferreira (2005), Giovannini (2004) and Guasseli (2005).

Research carried out in the Taim wetland showed a tendency for the number of aquatic macrophyte species to decrease with increasing water depth. If the water level is kept high for too long, the floating species tend to dominate the modified environment, thus altering the macrophyte communities, also leading to changes in the habitats of other species, particularly the communities of aquatic birds and mammals (MOTTA-MARQUES; IRGANG; GIOVANNINI, 1997).

The site is also rich in birds, with more than 230 species, 78 of which are waterfowl, representing 18 families. Many of these species have a migratory habit and do not breed at this latitude. Nine species come from the Northern Hemisphere and remain in RS in the warmer months, and five species originate from the Southern Hemisphere and occur in the colder months (BURGER; RAMOS, 2007).

This number corresponds to approximately 34.7% of all bird species found in the state (BENCKE *et al.,* 2010) and 14.8% of all existing waterfowl species. As a result, Rio Grande do Sul represents one of the most diverse areas of waterfowl in Brazil, along with the Pantanal-Matogrossense, as this region is part of the migratory routes of species from the northern and southern hemispheres.

As for the rest of the fauna, there are still few studies carried out in the Taim ESEC, with 63 species of fish having been recorded so far (GARCIA *et al.*, 2006; CORREA *et al.*, 2011), 18 species of amphibians (GAYER; KRAUSE; GOMES, 1988), 21 species of reptiles (GOMES; KRAUSE, 1982), and it is thought that there are more than 40 species of mammals (MMA, 2013).

In studies carried out at the Taim ecological station by different researchers, it has been reported that water levels have a clear effect on bird populations, causing a reduction in species richness. Mahler, Kindel and Kindel (1996) reported that in the spring of 1991, when water levels were high and some environments were submerged, there was a lower frequency of records of species such as *Pluvialis dominica*, *Calidris* sp. and *Charadrius* sp.

In the study carried out by Melo (1990), between January 1986 and April 1989, during a period of drought, there was a report of a reduction in aquatic habitats, and only resident species such as *Cygnus melancoryphus*, *Fulica* sp., and *Coscoroba coscoroba* could be found. The research into the sociability and territoriality of *Chauna torquata* (tacha) carried out by Valle (1991) during the same period of drought (1989 to August 1990), describes that there were few activities related to the reproduction of this species, and the finding that the reproductive period of *C. torquata* is directly influenced by the level of water and the availability of food.

## 8. MATERIALS AND METHODS

The organization of the processes adopted in this research is first presented by means of a flowchart, shown in figure 5. Subsequently, there is a detailed elucidation of the system for each stage.

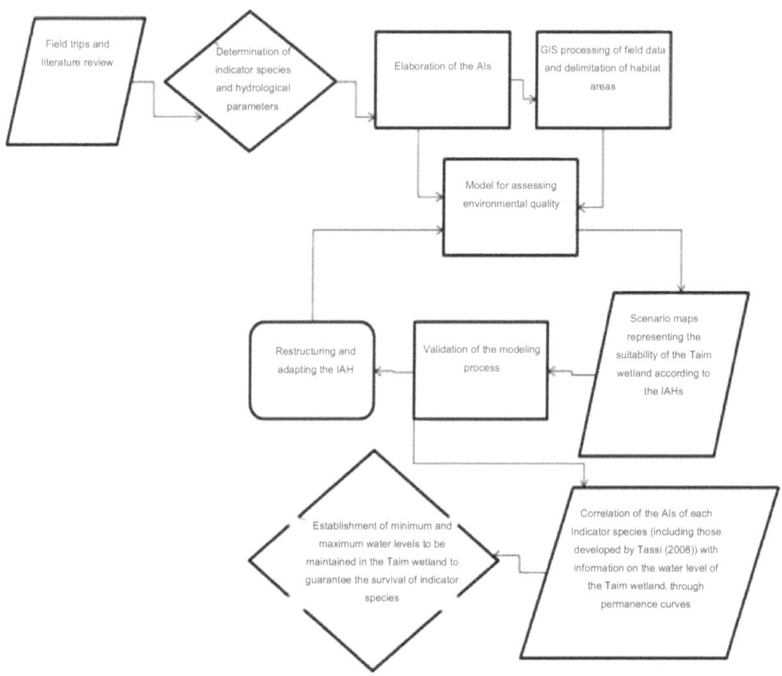


Figure 5 - Flowchart summarizing the processes used in the research methodology.

### 8.1 Determination of bioindicators

In order to determine indicator species for the Banhado do Taim, the choice was made of emergent aquatic macrophyte plant species and representatives of avifauna, herpetofauna and mastofauna. Emergent aquatic macrophytes are characterized by being more affected by variations in water levels than floating aquatic macrophytes, showing lower tolerance limits.

The use of vertebrates as environmental indicators has been used to evaluate different ecosystems and is justified by the ease with which individuals can be visualized and identified, the fact that populations are more stable, and the fact that it requires less time and effort to develop management strategies (BAKER; SCHONEWALD-COX, 1986; ODUM, 2004).

The processes used to define the indicator species of Banhado do Taim, as well as the recognition of favorable environmental conditions and the distribution and location of populations in the wetland environment are shown in Table 1.

<table>
<tr><th>Process</th><th>Material</th><th>Determination of indicator species</th></tr>
<tr><td>Recognizing the biodiversity of Banhado do<br>Taim</td><td rowspan="3">Bibliographic review of studies and research carried out in the ESEC Taim region and field observations</td><td rowspan="3">Emergent aquatic macrophytes, herpetofauna, avinauna and mastofauna spotted in the field, which had information on the hydrological conditions necessary to guarantee their survival.<br>They have the characteristics described by Odum (2004) - small tolerance limits (to changes in water level); sensitivity to small environmental changes and easy identification.</td></tr>
<tr><td>Obtaining information on hydrological conditions favorable to the survival of species in Banhado do Taim (tolerable hydrological condition ranges, water level height)</td></tr>
<tr><td>Information on the distribution of species in the Banhado do Taim area</td></tr>
</table>

Table 1 - Processes and materials used to define the indicator species for Banhado do Taim

## 8.2 Field trips and habitat mapping

The field trips took place between February and April 2014, with one in each month. The specimens were observed by walking along the BR-471 highway (a 16 km stretch, as shown in Figure 6), which borders the ESEC Taim, with the aid of binoculars and the collection of geographic location information (coordinates) using a Garmin GPS (Global Positioning System) receiver (eTrex® 10), photographic recording and filming with a digital camera (Sony CyberShot DSC-HX1 Semi Professional), as well as describing the locations observed.

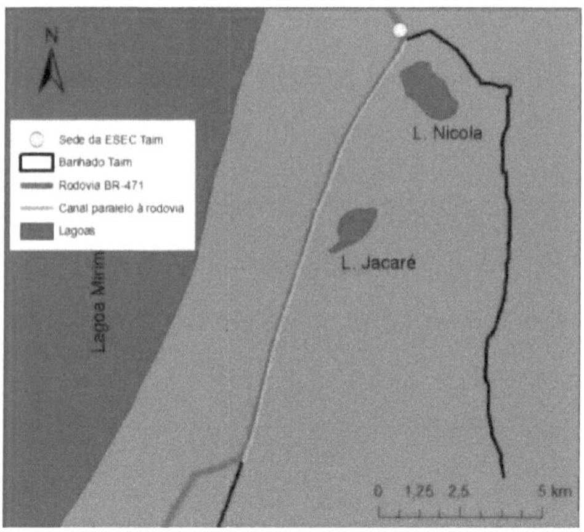


Figure 6 - Representation of the 16 km stretch of the BR-471 highway that cuts through ESEC Taim, along
which we traveled on foot and observed the marshland environment.

While traveling along the highway (a 16 km stretch of the BR-471), whenever a potential species was spotted (with the aid of binoculars with a 7x35 viewing range), the location of the point was marked on the Global Positioning System (GPS), the distance between the species and the point marked on the highway was measured, and the environment was photographed and described, as illustrated in figure 7.

Figure 7 - Example of the process of traveling along the BR-471 highway and observing the species in the interior of Banhado do Taim.

The field spreadsheet also recorded information on the sampling date, the season, and the hydrological conditions for the corresponding hydroperiod, such as the water level by reading the rulers. This information was then processed using Geographic Information Systems software, ArcMap 10.0 and IDRISI Selva 17.0, in order to make it compatible with the environmental quality assessment model developed specifically for Banhado do Taim by Tassi (2008).

The product of the data survey, together with information on the distribution of the species in the area, obtained from studies carried out at ESEC Taim, helped to delimit the areas considered to be habitats within the marsh for the species suitable to be adopted as indicators in this study.

These habitat areas were then mapped and represented using polygons in the ArcMap software, and converted into a "raster" format generated using the IDRISI Selva software, with specific formatting compatible with the model for assessing environmental quality, with 316 lines and 141 columns.

## 8.3 Elaboration of the AIs

For each indicator species, a habitat suitability index was drawn up, a process already explained and exemplified in section 6 of this study, using one of the criteria proposed in table 2.

Table 2 - Variables used to calculate the AIs

| IAH | | Variable |
|---|---|---|
| - 0 (unfavorable environment for the indicator); | | - the height of the water table needed to guarantee the conservation of the species; |
| | X | |
| - 1 (totally suitable environment for the indicator). | | - The area needed to be kept free from flooding during the reproductive period. |

The AIs were later integrated into the environmental quality assessment model (Figure 8) developed by Tassi (2008). This was developed in the FORTRAN programming language. This model allows the habitat suitability conditions to be assessed on a monthly basis in the BT.

To evaluate the IAHs, this study used the historical record of water levels in the BT, considering the premise that indicator species are the result of hydrological conditions observed in the past. Therefore, their maintenance can be ensured if the same behavior is reproduced in the future. To do this, we used the results of a 44-year period (Jan-1960 to Dec-2003) of hydrological data on monthly water levels, obtained from field data and hydrological-hydrodynamic modeling (TASSI, 2008).

In the hydrodynamic model, the BT is represented by dividing it into 41 distinct cells (Figure 8a), as described in Villanueva (1997), where you can also find more details about the model, the calibration and validation process. The data required for input are rainfall, evaporation, the amount of water collected for irrigation and the water level of Lagoa Mirim.

For the environmental quality assessment model, the water level is obtained for each pixel (Figure 8c), based on the hydrodynamic modeling of cells in the centroid of each cell and the TIN (Triangulated Irregular Network) interpolator was also used for its spatialization in the BT. Next, to determine the water level, the image of the spatialized water level (Figure 8c) is subtracted from the background topography of the BT (Figure 8b), and the image with the spatialization of the water level in the BT is obtained (Figure 8d).

After obtaining the spatialized water level, the model relates it to the specific AHI of each species under analysis, producing a new image (Figure 8e) that expresses the suitability of the habitat on a scale of numbers and colors, with 0 meaning an unfavorable environment and 1 a totally favorable environment determined by its AHI, depending on the hydrological variable.

Thus, for each indicator species used in the study, 528 environmental suitability images were generated (similar to figure 8e), derived from the hydrological conditions for each month (from January to December) between the years 1960 to 2003. These images have a pixel resolution of 101.86 x 101.86 m, where for each pixel there is a determination of its suitability for the given species.

Using the same model, information was obtained such as average values and conditions in specific regions.

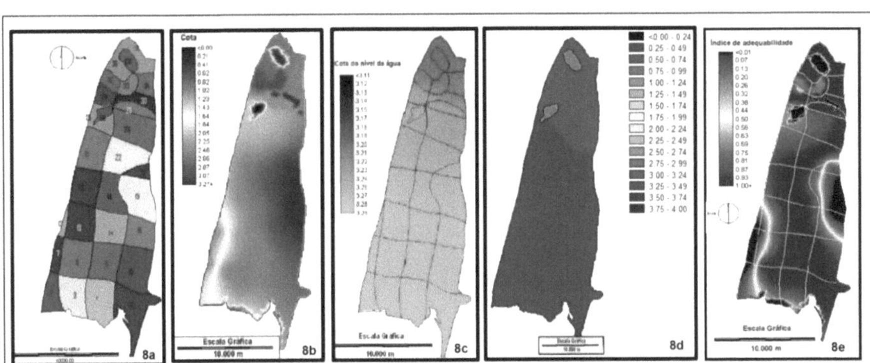


Figure 8 - Representation of the model for assessing environmental quality. Figure 8a - Hydrodynamic model. Figure 8b - Numerical Terrain Model (NTM). Figure 8c - Water level elevation. Figure 8d - Water level height. Figure 8e - Product of the modeling process, image of environmental suitability. Source: Adapted from Tassi, 2008.

A new version of the environmental quality assessment model (which was programmed solely to return information on habitat suitability as a function of the height of the water table in the wetland) was also created, adapted to provide information on flood-free areas during the simulations - for specific indicators, as will be discussed in the results.

In this case, instead of indicating the value of the AHI in the specialized image, images and outputs were generated that express the areas that would be suitable for a given species under a specific hydrological condition (1) and unsuitable locations (0), as well as an indication of the percentage of area under these conditions.

Thus, the environmental quality assessment model was combined with the 44-year historical series of monthly water level data from the BT, allowing information on the tolerance limits for the height of the water level or the area to be kept in appropriate conditions (free from flooding), defined for each species in its AHI, to be assessed.

## 8.4 Validation of AIs

As a way of improving the ecological niche modeling process using the AHI, the modeling was validated by checking the information from the AHI, the distribution of the species in the environment (representation of habitats) and the results of the images generated by the environmental quality assessment model, after modeling 44 years of data, by selecting the images generated for the indicator species using the environmental quality assessment model, as shown in Table 2.

| Images |
| --- |
| Drought period - Water levels below the average observed in the data series. |
| Flood period - Water levels above the average observed in the data series. |
| Period with BT water levels at the average level of the data series. |
| Average LV suitability over the entire period of the data series. |

Chart 2 - Images selected to validate the AIs.

The images analyzed were taken between the months of September and March, which coincide with the period of increased evapotranspiration and greater water withdrawals for irrigation, as well as corresponding to the reproductive phase of many of the species and the occurrence of migratory birds, which come to the region in search of food.

## 8.5 IAH restructuring and revalidation

In cases where the images generated through the modeling process were not satisfactory for a given indicator, there was a process of restructuring and readjusting the IAH according to the reality of Banhado do Taim. This process was applied when the regions mapped as occurring the species, under hydrological conditions considered suitable for the indicator (information available in the data series), showed suitability equal to 0 or very close to this value.

As a way of adjusting the AHI, the values relating to the tolerance limits supported by the indicator were adapted to the hydrological conditions of the wetland environment in question, always respecting the values indicated in the literature and field observations.

Once the IAH had been restructured, it was used again in the simulation with the 44 years of data, and a new evaluation of the spatialized product was carried out by analyzing the images obtained.

## 8.6 Drawing up permanence curves

In order to help manage the water resources of the Banhado do Taim region, we sought to establish minimum and maximum water levels to be maintained in order to guarantee the survival of the indicator species and consequently of the entire community associated with these organisms.

Since the hydrological management of the BT is carried out by means of rulers and linigraphs, the hydrological information is related to an elevation referenced in the IBGE system, so the information corresponding to the water level heights was related to a water level elevation value.

As a way of evaluating the water levels, permanence curves were drawn up to represent the cumulative frequency curve of the time series of water level levels, by determining the percentage of time that a given water level value has been equaled or exceeded in Banhado do Taim, as well as the average water level.

In the specific case of this study, permanence curves corresponding to the bioindicator organisms were generated, based solely on the information on water levels (contained in the cells of the hydrodynamic model) in the region of the habitat area of the indicators that had their AHI constructed as a function of the water level. The period considered in the analysis was between September and March of the historical series between January 1960 and December 2003 (the same period covered by the validation of the images).

In order to obtain a minimum and maximum water level value for the indicators, we analyzed situations that would not be totally ideal, but would ensure part of the environmental conditions necessary for the species, using the criteria organized in Table 3.

| Minimum and maximum water level | |
|---|---|
| Images generated through the environmental quality modeling process of drought and flood periods | AHI values between 0.12 and 0.50 |

Table 3 - Determination of the minimum and maximum water level values for the bioindicators.

With the period determined (month and year), a query was made in the table containing the series of information on the LV levels, separated by each cell of the hydrodynamic model. In this way, it was possible to select the level information for the indicator's habitat areas, under the predetermined conditions. In order to establish an average level, the analysis was based on a year with a water level closest to the average level (2.90 m), which showed good suitability for the bioindicator.

The series of water level data used as a reference for Banhado do Taim was cell 25 of the hydrodynamic model, as already established as a convention in other work carried out at ESEC Taim, since this is the location of a linigraph around Jacaré Lagoon. As in the assessment of the suitability images, the period assessed corresponds to the crucial period for the indicator species, spring and summer.

## 8.7 Proposed minimum and maximum water levels for Banhado do Taim based on indicators

In order to compare the seasonality of the levels in Banhado do Taim with the conditions that would sustain an environment favorable to all the species analyzed, the pre-determined data of minimum and maximum levels for each bioindicator were merged and compared with the BT's annual hydroperiod. The information on the bioindicators used in Tassi's research (2008) was also included in this process of analyzing the seasonality of the wetland's hydrological regime and in drawing up the proposal for an ecological level to be maintained in Banhado do Taim.

Finally, an analysis is made of the series of water levels in Banhado do Taim, in order to identify the natural oscillations of the water levels in the interior of the marsh over time and their relationship with the indicators.

# 9. RESULTS AND DISCUSSION

The results obtained through this research were organized following the sequence presented in item 8 of this study, Materials and Methods.

## 9.1 Determination of biondicators

The field trips provided a better understanding of the dynamics of the Banhado do Taim ecosystem and the identification of the different habitats and diversity of species that make up the environment. As for the aquatic vegetation, it was possible to see that the vegetation of the channel parallel to the highway is mainly characterized by the presence of floating macrophytes, unlike the vegetation of the marsh region, where there is a predominance of emergent macrophytes, with the species *Zizaniopsis bonariensis* and *Scirpus californicus standing* out for their abundance.

As for the fauna observed, there was a great diversity of birds, with around 40 species sighted, represented by 10 Orders, the most representative being Pelecaniformes, Charadriiformes, Anseriformes and Gruiformes, belonging to 14 different Families, including Ardeidae, Threskiornithidae, Rallidae, Anatidae and Ciconiidae. In addition, 3 species of reptiles (*Trachemys dorbigni*, *Phrynops hilarii* and *Caiman latirostris*) and 3 species of mammals (*Hydrochoerus hydrochaeris*, *Myocastor coypus* and *Lontra longicaudis*) were spotted. The following table (Table 3) illustrates some of the most significant species of birds associated with wetland environments observed in Banhado do Taim during the trip along the BR-471 highway.

Table 3 - List of the main bird species associated with the wetland environment observed during field trips to Banhado do Taim.

| Common name | Scientific name | Family | Order |
|---|---|---|---|
| Irerê | *Dendrocygna viduata* | Anatidae | Anseriformes |
| Coscoroba | *Coscoroba coscoroba* | Anatidae | Anseriformes |
| Tachâ | *Chauna torquata* | Anhimidae | Anseriformes |
| Crane | *Ardea cocoi* | Ardeidae | Pelecaniformes |
| Great Egret | *Ardea alba* | Ardeidae | Pelecaniformes |

| | | | |
|---|---|---|---|
| Heron | *Bubulcus ibis* | Ardeidae | Pelecaniformes |
| Little egret | *Egretta thula* | Ardeidae | Pelecaniformes |
| Punch | *Butorides striata* | Ardeidae | Pelecaniformes |
| Blackbird | *Vanellus chilensis* | Charadriidae | Charadriiformes |
| Joâo-grande | *Ciconia maguari* | Ciconiidae | Ciconiiformes |
| Dry-headed | *Mycteria americana* | Ciconiidae | Ciconiiformes |
| Jaçanâ | *Jacanajacana* | *Jacanidae* | *Charadriiformes* |
| Bigua | *Phalacrocorax brasilianus* | Phalacrocoracidae | Suliformes |
| Marsh harrier | *Pardirallus sanguinolentus* | Rallidae | Gruiformes |
| Yellow Carqueja | *Fulica leucoptera* | Rallidae | Gruiformes |
| Common waterfowl | *Gallinula galeata* | Rallidae | Gruiformes |
| White-backed stilt | *Himantopus melanurus* | Recurvirostridae | Charadriiformes |
| Spoon | *Platalea ajaja* | Threskiornithidae | Pelecaniformes |
| Black Sandpiper | *Plegadis chihi* | Threskiornithidae | Pelecaniformes |

Due to the high richness of bird species observed, it was decided to select those that seek food resources essentially in the aquatic environment or wet soil. Priority was therefore given to those most sensitive and restrictive to variations in water levels, thus disregarding larger heron species from the modeling process of this study, such as the cattle egret (*Ardea cocoi*) and great egret (*Ardea alba* ) and birds that use floating vegetation as habitat or refuge, such as the *jacana* (*Jacana jacana)*, marsh harrier (*Pardirallus sanguinolentus) and the* common moorhen (*Gallinula galeata*).

Following the pre-established criteria, the emerging aquatic macrophyte *Scirpus californicus*, the birds *Plegadis chihi and Mycteria americana*, as well as the reptiles *Caiman latirostris* and *Trachemys dorbigni* were determined as indicator species for this study.

The groups of bioindicators used in this study to develop HIAs, an emerging aquatic macrophyte plant from the Cyperaceae family, birds from the Threskiornithidae and Ciconiidae families and reptiles from the Emydidae and Alligatoridae families, have served as the basis for research into ecological niche modeling in different ecosystems, including the development of HIAs for wetlands, due to their specific characteristics and adaptations to these environments.

The experts who developed the indices for the Everglades region (USA) used AIs for different parameters related to reproduction, nesting, foraging and predation, such as the percentage of wetland available, the depth of the body of water, the percentage of area flooded during nesting to deter predators and the distance from a permanent body of water. Generating more complex and complete composite AIs for each bioindicator. This was not possible in this study due to the lack of information on the species, so the AIs were generated with reference to just one piece of information.

The IAHS for the species *S. californicus*, *P. chihi*, *M. americana* and *C. latirostris* were built using information on the height of the water depth needed to guarantee the conservation of the species, through bibliographic information and field data on the appropriate depth for its occurrence, ensuring its chemical processes, food availability and reproductive success.

For the *T. dorbigni* species, the information used to develop the AHI was the area needed to be kept free from flooding during the reproductive period, between October and January, guaranteeing greater protection for the nests, ensuring that they are free from flooding.

### 9.1.1 *Scirpus californicus* (Juss, 1789)

The bulrush (Figure 9) is a plant of the Cyperaceae family, and is characterized according to Kissmann, (1997). It has a simple, erect stem, usually between 0.60 and 2.00 m high, which at flowering time has a cluster of spikelets hanging from the apical part, above which is an acute bract, which appears to be a continuation of the stem.

Figure 9 - Specimens of *Scirpus californicus*. Source: UNICAMP (2014).

The reed has a trigonous stem with rounded angles, about 1 cm in diameter at the basal part; smooth, glabrous surface, dark green in color. Inside, the stems are filled with a spongy tissue. The stems can be continuously bent by the wind or the current of the water without being damaged.

The leaves are reduced to brown membranous sheaths, which cover the stem up to a height of between 0.20 and 0.30m. The upper sheath sometimes has a rudimentary blade. In the apical part of the stem there are pseudo-panicles (compound umbels), flat and pendulous, formed by rays of unequal length, the largest about 10 cm long, usually branched. In a well-developed plant there can be up to 150 short pedicellate spikelets, either isolated or in groups of two or three. Reed is a perennial plant that reproduces by seed. Additional stems develop from robust rhizomes, so that one plant forms a considerable mass of vegetation.

It is characterized by its origin on the American continent, where it also has the widest distribution, occurring from the west and south of the United States to Patagonia; in Brazil it is very common in flooded areas. The suitable environment for its development is wet lowlands, which can be temporarily flooded, where, as an emergent aquatic macrophyte, they are rooted to the bottom, partially submerged and partially out of the water. Flowering occurs in spring and summer in the south and southeast of Brazil.

### 9.1.2 *Plegadis chihi* (Vieillot, 1817)

The black curlew (Figure 10), also called tapicuru or white-faced curlew, is a Pelecaniform wader from the Threskiornithidae family, with an average height of 53 cm, chocolate-brown plumage with dark violet-green wings and tail, dark beak and legs and a red iris. The bird's beak is curved and between 15 cm and 18 cm long, adapted for submerging in water or mud. It occurs in areas of flooded fields, agricultural fields and wetlands with low emergent vegetation, usually in flocks (DARK-SMILEY; KEINATH, 2003; RYDER; MANRY, 1994; SICK, 2000; TROST, 1989).

Figure 10 - Adult specimen of *Plegadis chihi.* Source: Souza (2011).

Black curlews feed mainly on aquatic or wet soil invertebrates, especially insects (Orthoptera, Odonata, Hemiptera, Coleoptera and Diptera), earthworms, leeches, snails, spiders, crustaceans and fish (DARK-SMILEY; KEINATH, 2003). These resources are obtained in shallow water areas, marshes and estuaries, flooded or saturated fields and fodder or grain crops (including exposed/flooded soil), and they use touch as their main foraging strategy. Their long legs, elongated neck and curved beak are used to facilitate foraging in shallow waters and wet soils (KUSHLAN; BILDSTEIN, 1992; DARK-SMILEY; KEINATH, 2003; PINHEIRO; SILVA, 2013).

In the northern hemisphere, it is distributed between the United States and Mexico, moving locally in large flocks depending on the location of nesting sites, food and water levels. In South America, they occur mainly in Bolivia, Paraguay, Chile, Argentina, Uruguay and Brazil (DARK-SMILEY; KEINATH,

2003, RYDER; MANRY, 1994).

It is considered to be the most abundant non-migratory waterfowl species with the greatest biomass in the southern coastal region of Rio Grande do Sul. It rests on the ground and breeds in the vertical vegetation of wetlands during the warm season, usually between November and January (PINHEIRO; SILVA, 2013; SOUZA, 2011). Safran *et al.* (2000) highlight the regions close to emergent vegetation as an important foraging area, where there can be a greater abundance of prey, making it essential to maintain this type of vegetation in freshwater wetlands.

### 9.1.3 *Mycteria americana* (Linnaeus, 1758)

The dry-headed godwit (Figure 11) is a bird from the Order Ciconiiformes, family Ciconiidae, a resident warm-season, gregarious visitor, which is quite abundant in the region of the Southern Coastal Plain of Rio Grande do Sul. It is characterized by the absence of feathers on the head and neck, white body plumage, a black beak, legs, rémiges and retrices. They can measure around 95 cm in height, have a wingspan of 1.5 m and weigh between 2.00 and 3.00 kg (LOPES, 2006; VAN METER, 1985; SICK, 2000).

Figure 11 - Specimen of *Mycteria americana* recorded during a field trip to ESEC Taim on February 20, 2014.

This species feeds mainly on fish, at restricted depths (0.15 and 0.50 m) and, like *P. chihi*, its foraging strategy is "tactile probing", with selectivity of fast-moving prey through touch, depending on the perceptual touch of the beak and not just its visual capacity (DEPKIN; COULTER; BRYAN JR, 1992; VAN

METER, 1985; SICK, 2000)

They can be found in tropical and subtropical wetlands in regions characterized by seasonal fluctuations in water levels. The bird prefers wetlands and flooded fields, as well as areas where rice is planted during the flooding, vegetative and harvesting phases, and can form flocks of hundreds of individuals (PINHEIRO; SILVA, 2013; VAN METER, 1985). It is distributed from the southeastern United States to the northern region of Argentina (SICK, 1997).

### 9.1.4 *Caiman latirostris* (Daudin, 1802)

The yellow-bellied caiman (Figure 12) belongs to the Alligatoridae family and is the only species of Crocodilia that occurs in the state of Rio Grande do Sul. It is a medium-sized species of caiman, which can reach up to 3.5 m in total length, but with rare specimens larger than 2.0 m (ROSS, 1998). It is characterized by its uniform dark olive-green or greyish dorsal coloration, with black transverse stripes and a yellowish belly. The chicks and juveniles show a variation in dorsal coloration consisting of irregular dark transverse stripes that contrast with the lighter, yellow-green background (BORGES-MARTINS, *et al.*, 2007; CARREIRA; MENEGHEL; ACHAVAL, 2005).

Figure 12 - Specimen of *Caiman latirostris* recorded during a field trip to ESEC Taim, on February 17, 2014.
February, 2014.

Its distribution in Brazil ranges from Rio Grande do Norte to Rio Grande do Sul, along the coasts of Espirito Santo, Rio de Janeiro, São Paulo and Santa Catarina, as well as in northeastern Argentina, southeastern Bolivia, Paraguay, northern Uruguay and eastern Brazil (CARREIRA; MENEGHEL; ACHAVAL, 2005; ROSS, 1998).

*C. latirostris* generally occurs in lentic environments and is often found in marginal lagoons, freshwater and saltwater marshes, swamps, mangroves and rivers, as well as occupying small temporary bodies of water, such as pools (BORGES-MARTINS *et al.*, 2007; MOULTON; MAGNUSSON; MELO, 1999). An opportunistic species, it feeds mainly on arthropods, molluscs (*Ampullaria* sp.) and small vertebrates (MELO, 1990).

The females build nests in decaying plant matter, on the banks of bodies of water or on floating aquatic vegetation (BORNSCHEIN *et al.*, 2012) between December and January, where they lay between 18 and 50 eggs. The chicks hatch around March (BORGES-MARTINS *et al.*, 2007; CARREIRA; MENEGHEL; ACHAVAL, 2005). The temperature during egg incubation plays a decisive role in determining the sex of the animal, where eggs incubated between 29°C and 31°C result in females, and eggs incubated at a temperature of around 33°C result in males. Parental care is provided by the females, who look after the young for a period of one to two years (CARREIRA; MENEGHEL; ACHAVAL, 2005).

### *9.1.5 Trachemys dorbigni* (Duméril and Bibron, 1835)

The tiger turtle (Figure 13) is a chelonian from the Emydidae family, characterized by a tall, oval to elongated carapace up to 0.25 m long, with yellowish-green coloring and black spots. There is great individual and ontogenetic variation in this color pattern. The plastron is yellow with a large central black spot, which extends along the sutures of the shields. The head has a series of colored longitudinal lines, green and yellow, with two yellowish or orange bands behind the eyes. The limbs are dark and also have a pattern of yellowish longitudinal lines (BORGES-MARTINS *et al.,* 2007; CABRERA, 1998; ROCHA, 2005).

Figure 13 - A specimen of *Trachemys dorbigni* found during a field trip to ESEC Taim, on the banks of highway BR-471, on February 20, 2014.

This species is diurnally active and can be easily observed when it spends long periods basking (exposed to the sun) on the banks of bodies of water, sometimes swimming on the surface with its head above water level, or walking in the fields or on the sand during oviposition periods (BORGES-MARTINS *et al.,* 2007; GOMES; KRAUSE,1982). It occurs in northeastern Argentina, Uruguay and southern Brazil (LEMA; FERREIRA, 1990). It is considered an opportunistic omnivorous animal, consuming a diversity of plant and animal items (HAHN, 2005).

The nests can be located in sandy, clayey and stony substrates, as well as along the sides of the road (BAGER, 2003). These sites can be flat or a few meters above the water level, have little or no vegetation cover and are permanently exposed to the sun. On average, 12 eggs are laid per nest, and a female can lay up to three eggs in a single reproductive season. The eggs are elliptical and the shell has a parchment-like consistency. Incubation lasts around 110 days and temperature is a determining factor in the incubation period and in determining the sex of the chicks (BAGER, 2003; KRAUSE; GOMES; LEYSER, 1982; MOLINA, GOMES, 1998).

## 9.2 Habitat mapping

Identification of the distribution of the indicator species in the Banhado do

Taim environment and delimitation of their habitat area occurred mainly through reference to research carried out at the site. This information was also supplemented by data obtained during field visits (Table 4). Figures 14 and 15 show the distance of each bioindicator in Banhado do Taim.

Table 4 - Information base for mapping the habitat area of the indicator species in Banhado do Taim

| Species | Basis |
|---|---|
| ***Scirpus californicus*** | Information contained in the research carried out by Guasseli (2005), which sought to portray the dynamics of the vegetation in Banhado do Taim, resulting in the representation of areas of occurrence of the main classes of macrophytes and complementation with field data. |
| ***Plegadis chihi*** | Data from field research, where it was possible to identify a strong relationship between the species and permanent bodies of water, such as the canal that runs parallel to the BR-471 highway and the banks of the Nicola and Jacaré lagoons. |
| ***Mycteria americana*** | Data from field research, with the area of greatest concentration of the bird being near the Jacaré Lagoon and the canal parallel to the BR-471 highway. |
| ***Caiman latirostris*** | This information is contained in the research carried out by Melo (1990), who considered the area of occurrence of the species to be the Jacaré channels, parallel to the BR-471 highway, the Jacaré Lagoon and the marshland region. Melo (1990) assumed that the distribution of caimans in the ESEC Taim is similar to that described for alligators by Taylor (1984)[1] , where adult males inhabit canals - deeper water regions - and females, marshland regions, with shallower water, dense vegetation and smaller home ranges. The study was also supplemented with field data. |
| ***Trachemys dorbigni*** | The information contained in the research carried out by Gomes and Krause (1982) indicates that chelonians are found in all the bodies of water present in the region, from small bodies of water that form on lower ground, in the channels that surround the marsh or interconnect lagoons, to large bodies of water. In the field, nests with already hatched eggs were spotted on the banks of the BR-471 highway, near the canal parallel to the road. |

I[1]TAYLOR, D. Management implications of an adult female alligator telemetry study. In.: Annual Conference of the Southeastern Association of Fish Wildlife Agencies.1984. **Proceedings of Conference**. v. 38, p. 221-227. 1984.

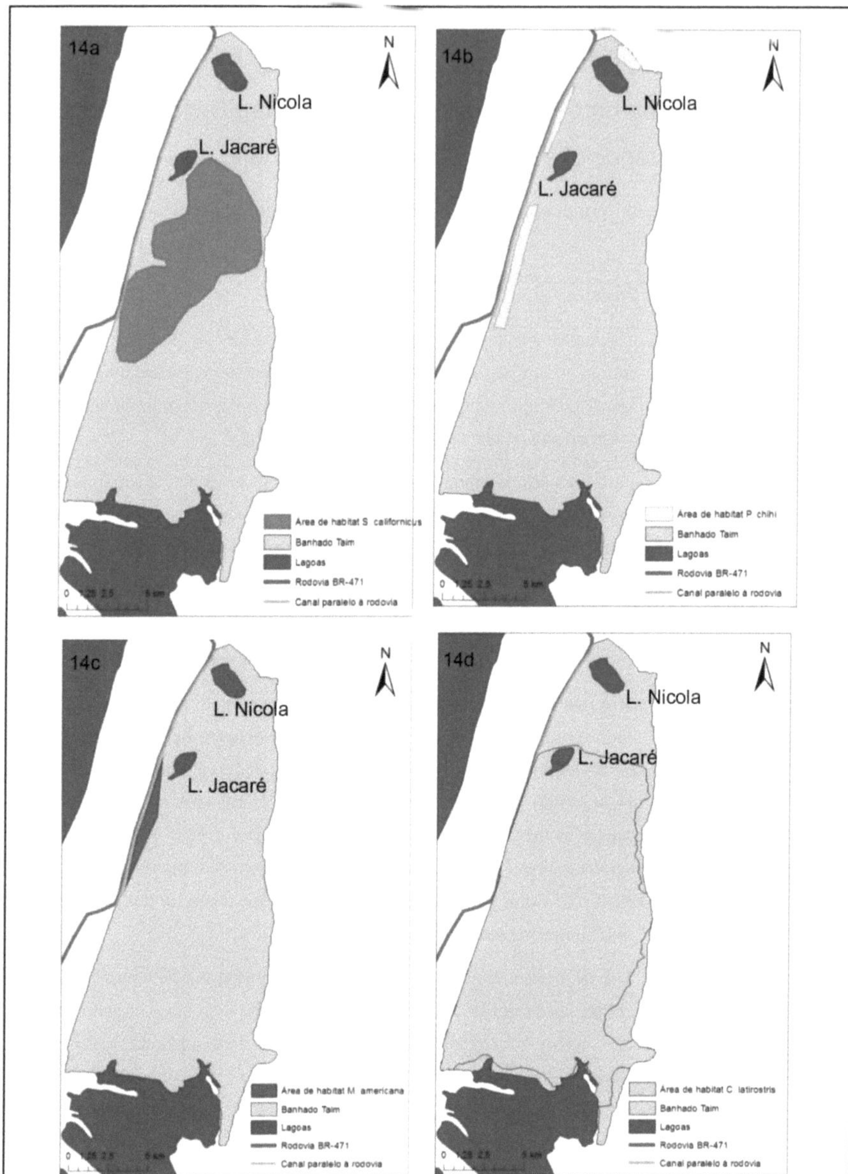


Figure 14 - Habitat mapping. Figure 14a - *Scirpus californicus.* Figure 14b - *Plegadis chihi.* Figure 14c - Mycteria *americana. Figure 14d - Caiman latirostris.*

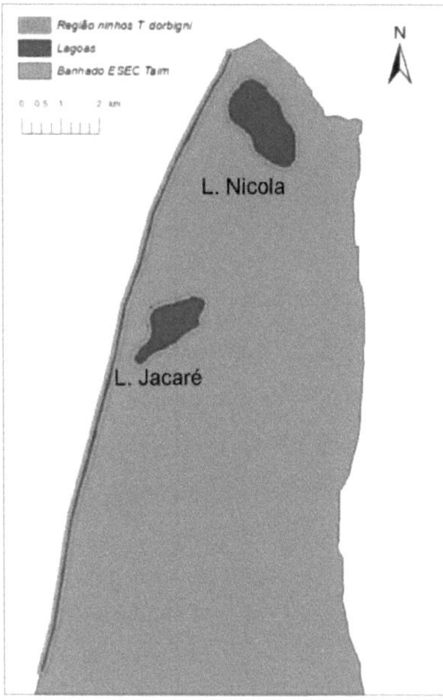


Figure 15 - Map of possible nesting areas for *Trachemys dorbigni* in the Banhado do Taim region.

## 9.3 Drawing up the AIs

By means of a bibliographic survey, eco-hydrological information was obtained on the water depth required for the species *S. californicus*, *P. chihi*, *M. americana* and *C. latirostris*. As well as information on the area to be kept free from flooding during the breeding season for *T. dorbigni* (Table 5). The AHI developed for each bioindicator is presented below.

Table 5 - Eco-hydrological information used in the process of drawing up the AIs.

| Indicator | Water blade height |
|---|---|
| *Scirpus californius* | Motta-Marques, Irgang and Giovannini (1997) - for herbaceous emergent plant species in Banhado do Taim, the ideal water level would be between 0.10 and 1.50 m. |
| | Florida Fish and Wildlife Conservation Commission (2002) - determines that the ideal average water level is between 0.60 and 0.91 m. |
| | Hydra Aquatic Inc. (2006) - ideal locations would be those with moist soil and a water table height of up to 0.91m. |

| *Plegadis chihi* | Kushlan and Bildstein (1992) - this species feeds in very shallow water, usually between 0.05 and 0.25 m deep. |
|---|---|
| | Field data (April 2014) - it was possible to check the height of the water level in two different places used by *P. chihi* when foraging in the vicinity of the Mangueira lagoon, where there was a variation between 0.08 and 0.11 m (information obtained by measuring with a ballpoint pen, photographic record and subsequent measurement with a ruler). |
| Mycteria *americana* | Coulter and Bryan (1993) - the foraging depth of *M. americana* varied between 0.04 and 0.63 m, but the birds were seen more frequently in places with a 0.25 m depth range. |
| | Depkin, Coulter and Bryan (1992) - preference for feeding sites between 0.15 and 0.50 m deep. |
| *Caiman latirostris* | Melo (1990) - captured individuals of the yellow-bellied caiman in the BT, and related the total length of the individuals to the depth of the body of water inhabited by the specimens. Most of these captures occurred at depths between 0.20 m and 1.20 m. |
| | Verdade, Lavorenti and Packer (1993) - reproductive management of yellow-bellied caimans in captivity, where the water depth of the ponds is kept to at least 1.00 m. |
| **Indicator** | **Area to be kept free of flooding during the breeding season** |
| *Trachemys dorbigni* | Bager (2003) - investigated the biology and reproductive ecology of *T. dorbigni* in the Banhado do Taim region and found that the species' spawning period begins in October and lasts until January. However, periods of high rainfall in early spring can interfere with this process and delay the nest-laying period. |
| | Bager and Rosado (2010) - defined a distance of terrestrial habitat to be preserved around bodies of water to ensure the protection and integrity of chelonian nests in ESEC Taim. For *T. dorbigni*, maintaining a distance of 94 m would guarantee protection from flooding for at least 90% of the nests. |

### 9.3.1 Habitat Suitability Index for *Scirpus californicus*

In the first instance, to develop the AHI for the species *S. californicus,* the information on the minimum water level corresponding to 0.10 m was used, in reference to the data collected by Motta-Marques, Irgang and Giovannini (1997), due to the fact that the research was carried out exactly in Banhado do Taim. As for the information on the ideal maximum water level for the species, we used that stipulated by the "Florida Fish and Wildlife

Conservation Commission" (2002) and "Hydra Aquatic Inc." (2006), which consider a maximum water level height of around 0.90m. This results in the AHI shown below (Figure 16).

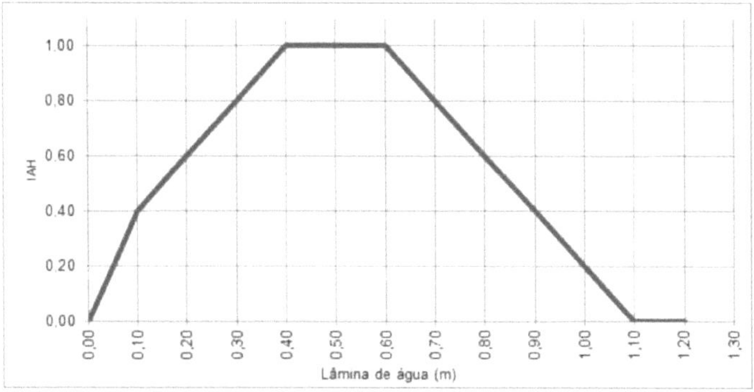


Figure 16 - First attempt to develop the Habitat Suitability Index for the species *Scirpus californicus* for Banhado do Taim.

The index seeks to portray the plant's tolerance levels to the seasonal water levels of the Banhado do Taim, considering dry sites with a water table lower than 0.00 m as not tolerable. A water table height of 0.05 m, which would characterize wet soil, was assumed to be a tolerable situation for the plant.

Information on the ideal or optimum water table height for the species was not found in the literature review, so it was assumed that swampy environments with water table heights between 0.40 m and 0.60 m were considered to be completely suitable for the species. This is similar to what was proposed by Tassi (2008) for *Z. bonariensis*, the most abundant aquatic macrophyte in the interior of the marsh and which is distributed in patches close to the *S. californicus* habitat area. Suitability for this species was considered to be maximum for water depths of 0.50 m.

The suitability of the environment would start to become unfavorable again the closer it got to 0.90 m, which is considered to be the maximum water level tolerated by the species. This threshold could compromise the plant's chemical and biological processes. As this value is approximate, a higher value was set for zero suitability, extending this to 1.10m.

## 9.3.2 Habitat Suitability Index for *Plegadis chihi*

Based on field observations, the information from Kushlan and Bildstein (1992) was used to develop the AHI for the *P. chihi* species. The resulting AHI for this species is shown in figure 17.

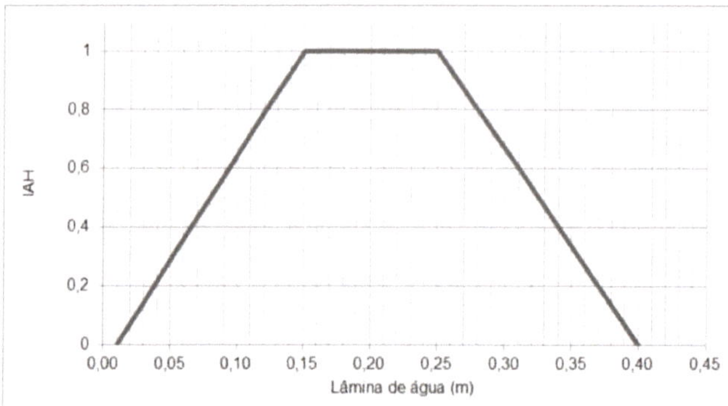


Figure 17 - Habitat suitability index developed for the species *Plegadis chihi* for Banhado do Taim.

The index shows that the bird has a very specific tolerance limit for the water level (up to 0.37 m), with its optimum being concentrated between 0.15 m and 0.25 m, which is strictly related to its foraging habits. As this species uses "tactile probing" to locate its prey - which refers to the use of the beak below the surface of the ground or the water blade - water depths greater than 0.25 m can hinder the capture of prey due to the length of the bird's beak, as well as making it difficult to move around.

Although *P. chihi* consumes a wide variety of prey, if individuals cannot find suitable foraging sites with plenty of food, they will move to places where food is more readily available (DARK-SMILEY; KEINATH, 2003). It would be very important to guarantee an abundance of food during the breeding season, which usually takes place between November and January, as Banhado do Taim is an important breeding site for this species, with around 12,000 breeding pairs having been identified in ESEC Taim in November 1980 (SICK, 1993).

## 9.3.3 Habitat Suitability Index for *Mycteria americana*

During the first field visit, in February 2014 (level on the rulers between 0.30 and 0.40 m), 22 individuals were observed in the BT, corroborating the range of values presented in the international literature used to compile the AHI for this species (Figure 18).

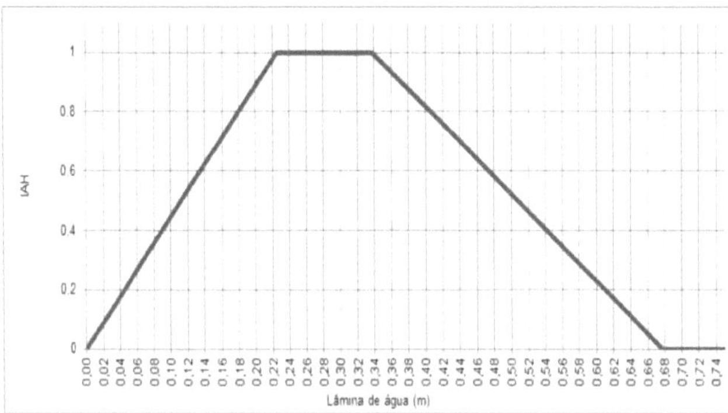


Figure 18 - Habitat suitability index developed for the species *Mycteria americana* for Banhado do Taim.

The index seeks to express mainly the foraging habits of the species, which uses prey that occurs at specific depths between 0.15 and 0.50 m and also due to its "tactile probing" strategy carried out with the beak. This foraging strategy requires an abundance of food, which at water depths of less than 0.10 m would result in a reduction in the number of prey, especially fish species. Water depths greater than 0.60 m would impair the feeding process of the individuals as it would be impossible to catch prey due to the length of the bird's beak.

In addition to feeding, *M. americana* populations have reproductive cycles that are synchronized with periodic fluctuations in water levels (GAWLIK *et al*., 2010; LO GALBO *et al.,* 2012). Intense changes in the hydroperiod in the Everglades region, USA, have led to a reduction in the population of this species (VAN METER, 1985), making it threatened with extinction in that country (U.S. FISH AND WILDLIFE SERVICE, 2014).

The bird makes migratory movements between breeding areas and feeding areas, and although there have been few studies on these movements, the main orientation is environmental conditions and food availability (COULTER *et al.*, 1999).

### 9.3.4 Habitat Suitability Index for *Caiman latirostris*

The combination of eco-hydrological information resulted in the AHI for the *C. latirostris* species, shown below (Figure 19).

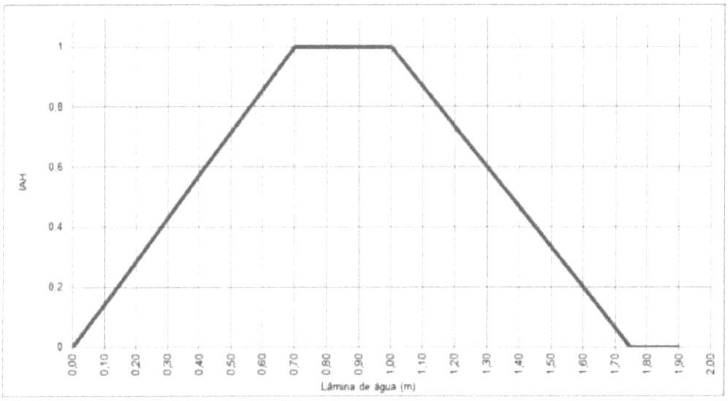


Figure 19 - Habitat Suitability Index developed for the species *Caiman latirostris*, for Banhado do Taim.

The index shows the height of the water level that would guarantee the survival of the species, especially for adult individuals, which need greater depth to move by swimming, since their movement on land is limited, not exceeding 200 m from a body of water (CARREIRA; MENEGHEL; ACHAVAL, 2005). Also according to Fett (2005), maintaining water levels higher than at least 0.60 m would facilitate the copulation process and guarantee a diversity of food resources for *C. latirostris*.

These principles corroborate the information used in the construction of the AHI, where it was proposed that maximum suitability for the yellow caiman would be maintained with a water level height of between 0.70 m and 1.00 m. Values below this level would impair adult movement in the aquatic environment. Values below these conditions would impair the movement of adult individuals in the aquatic environment and the copulation process. Values above 1.00 m would impair the copulation process, and depending on the body of water, could lead to flooding of the banks, reducing the space available for resting, thermoregulation and nest building.

### 9.3.5 Habitat Suitability Index for *Trachemys dorbigni*

To develop the AHI for this species, we used information from the research by Bager and Rosado (2010), carried out at ESEC Taim (Figure 20).

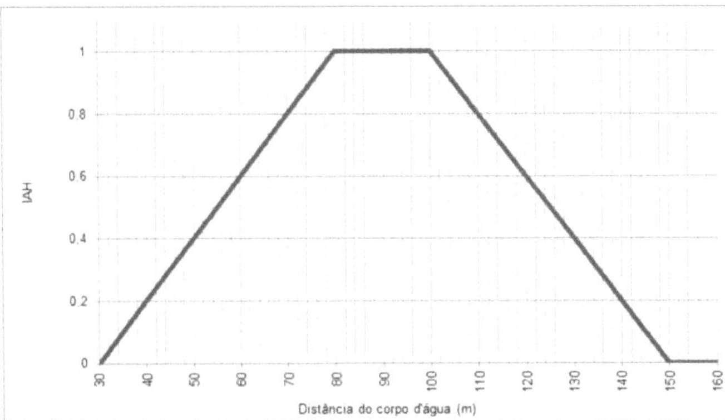


Figure 20 - Habitat Suitability Index developed for the species *Trachemys dorbigni* for Banhado do Taim.

Based on the information on the ideal distance to be kept free from flooding during the reproductive period, we tried to create a tolerance interval that included the 94 m information. Therefore, the maximum suitability, which would guarantee the conservation of the largest possible number of nests, would be between 80 m and 100 m from the body of water.

The selection of a spawning area involves various factors, including the guarantee of reproductive success by choosing areas where environmental impacts, predation and anthropogenic action are reduced (BAGER, 2003). A greater distance between the nests and the body of water reduces the risk of flooding. In the index, nests located at distances of less than 60 m were considered to be the most favorable areas for flooding.

However, nests located at long distances can also be considered a risk factor for the chicks, as they have to travel a greater distance to the nearest body of water, leaving them exposed to predators and climatic factors (BAGER, 2003). Thus, in the IAH, this maximum distance begins to be unfavorable after 100 m.

In the validation process, the images generated by the model for assessing the environmental quality of the Banhado do Taim are presented, which aim to reproduce the tolerance limits established in the IAHs of the indicator species. The color scale seeks to demonstrate the suitability of Banhado do Taim in different situations, as shown in Table 6.

Table 6 - Period selected for each indicator to assess environmental quality.

| **Indicator** | **Flood period** | **Drought period** | **Period with water levels at the average of the data series** | **Average BT suitability** |
|---|---|---|---|---|
| *S. californicus* | January 1985 (Figure 28a) | January 1989 (Figure 28b) | January 2001 (Figure 21a and Figure 28c) | Average corresponding to the entire data series (Figure 21b and Figure 28d) |
| *P. chihi* | December 1983 (Figure 22a) | December 1992 (Figure 22b) | January 2001 (Figure 22c) | Average corresponding to the entire data series (Figure 22d) |
| *M. americana* | February 1983 (Figure 23a) | February 1992 (Figure 23b) | February 2000 (Figure 23c) | Average corresponding to the months of September to March in the data series (Figure 23d) |
| *C. latirostris* | January 1985 (Figure 24a) | January 1989 (Figure 24b) | January 1960 (Figure 24c) | Average corresponding to the entire data series (Figure 24d) |
| *T. dorbigni* | December 1977 (Figure 25a) | January 1991 (Figure 25b) | December 1999 (Figure 25c) | Average corresponding to the months of September to March in the data series (Figure 25d) |

## 9.4.1 Validation of the IAH for *Scirpus californicus*

When analyzing a period with water levels comprised in the average of the data series (Figure 21a), it can be seen that within the area delimited as a habitat area, there is no expected suitability in relation to the height of the water level, with AHI values between 0.18 and 0.50 (orange and yellow). Ideal suitability was only found in the center of the patch, with values between 0.75 and 1.00 (dark green and blue). The average suitability (Figure 21b) was also unsatisfactory, with suitability ranging from 0.19 to 0.39 (orange). In view of these analyses, it was decided to restructure the AHI for the *S. californicus* species.

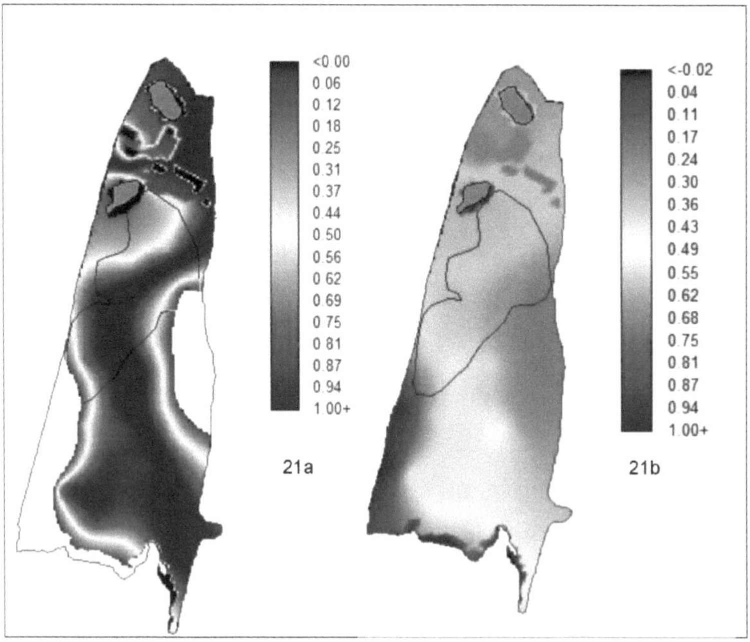


Figure 21 - Images generated through the environmental quality modeling process for *S. californicus*. Figure 21a - Period with water levels at the average of the data series. Figure 21b - Average BT suitability.

## 9.4.2 Validation of the AHI for *Plegadis chihi*

*The* period with water levels comprised of the average of the data series

(Figure 22c) illustrates an excellent suitability index ratio for the species' distribution patch located in the northern part of the BT, with values between 0.8 and 1 (green and blue coloration), predominantly blue. In the patch of the species' habitat located between the Nicola and Jacaré lagoons, there was a good IAH ratio, with values between 0.5 and 1 (yellow, dark green and blue). However, the patch located to the south of Jacaré lagoon and close to the channel parallel to the BR-471 highway was found to be an unfavorable area (suitability 0).

However, in dry conditions (Figure 22b), this was the habitat region that proved most suitable for the species, showing that depending on the hydrological condition there can be an alternation of habitat areas, mainly related to the availability of resources.

In a situation of flooding of the BT (Figure 22a), none of the habitat area patches proved to be favorable for the occurrence of the species (suitability 0), due to the fact that the bird requires a very restricted water level. The image representing the long-period average water level (Figure 22d) was also unsatisfactory for the species, with suitability ranging from 0.3 to 0.03 (orange color).

In addition, for the spatial representation of suitability for this species, it should be noted that there are major uncertainties associated with the bottom topography of Banhado do Taim in the region of the channel parallel to BR-471, precisely where most of the individuals of *P. chihi* were found. This may have hampered the representation of this species, but it is hoped that in the future a more detailed topographic survey will help in the process of revising the index for this species.

When analyzing the suitability of the BT, disregarding only the region delimited as the habitat area of *P. chihi*, it is possible to see that there are very favorable regions for the occurrence of the species. As these are regions located in the interior of the marsh, which are difficult to access, it was not possible to verify the individuals *in loco*.

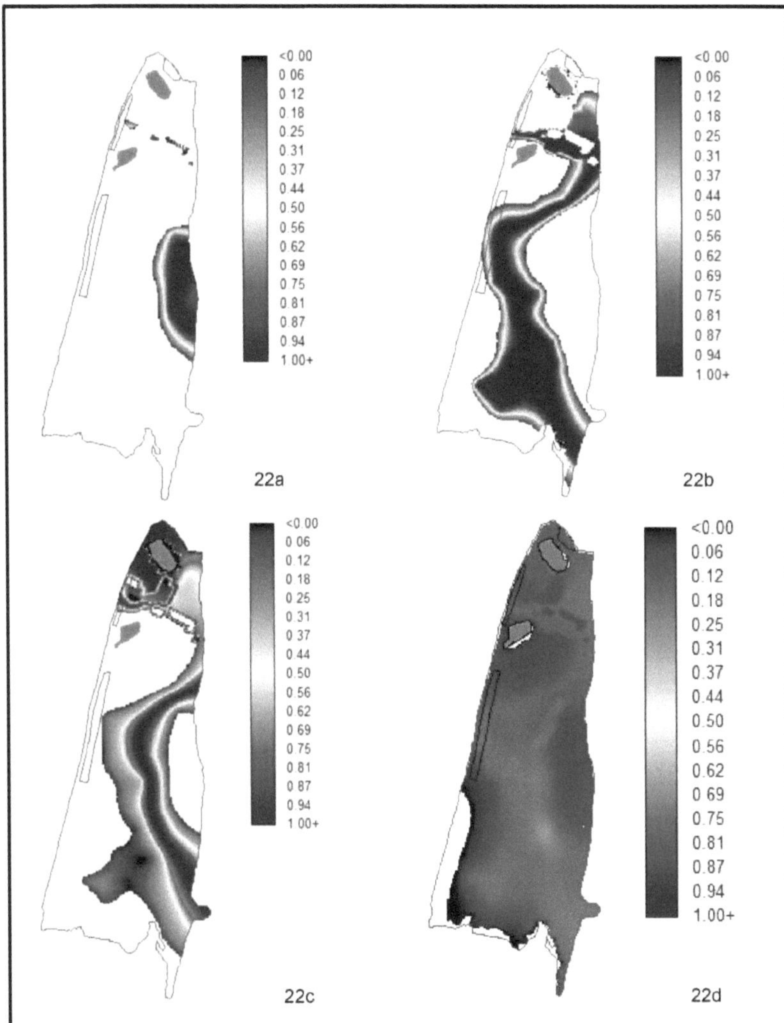


Figure 22 - Images generated through the environmental quality modeling process for *P. chihi*. Figure 22a - Flood period. Figure 22b - Drought period.

Figure 22c - Period with water levels at the average of the data series.

Figure 22d - Average BT suitability.

When analyzing a period with water levels comprised of the average of the data series on BT water levels (Figure 23c), it is possible to see a good relationship between the index and the distribution of the species, showing suitability between 0.5 and 1, especially in the area near Jacaré Lagoon (colored blue). In the dry season image (Figure 23b), the region near Jacaré Lagoon remains the most favorable for the species' foraging activities, with suitability between 0.2 and 0.6 (orange, yellow and light green).

During the flood period (Figure 23a), there were no sites favorable to the species (suitability 0) in the distribution area of the individuals observed, due to the fact that the bird requires a very restricted water level height, occupying sites of around 0.25m.

The image representing the average suitability of the BT (Figure 23d) was not very satisfactory for *M. americana*, showing suitability between 0.12 and 0.24 (orange color). It should be noted that the inaccuracy of the representation of the bottom topography of the BT in the region close to the channel parallel to BR-471 should be taken into account, as already discussed for *P. chihi*.

When analyzing the suitability of the BT, disregarding only the region delimited as a habitat area for *M. americana,* it is possible to see that there are several favorable regions for the occurrence of the species. Many of these are located in the interior of the marsh, which makes access difficult and also makes it difficult to see the specimens, due to the dense aquatic vegetation that emerges.

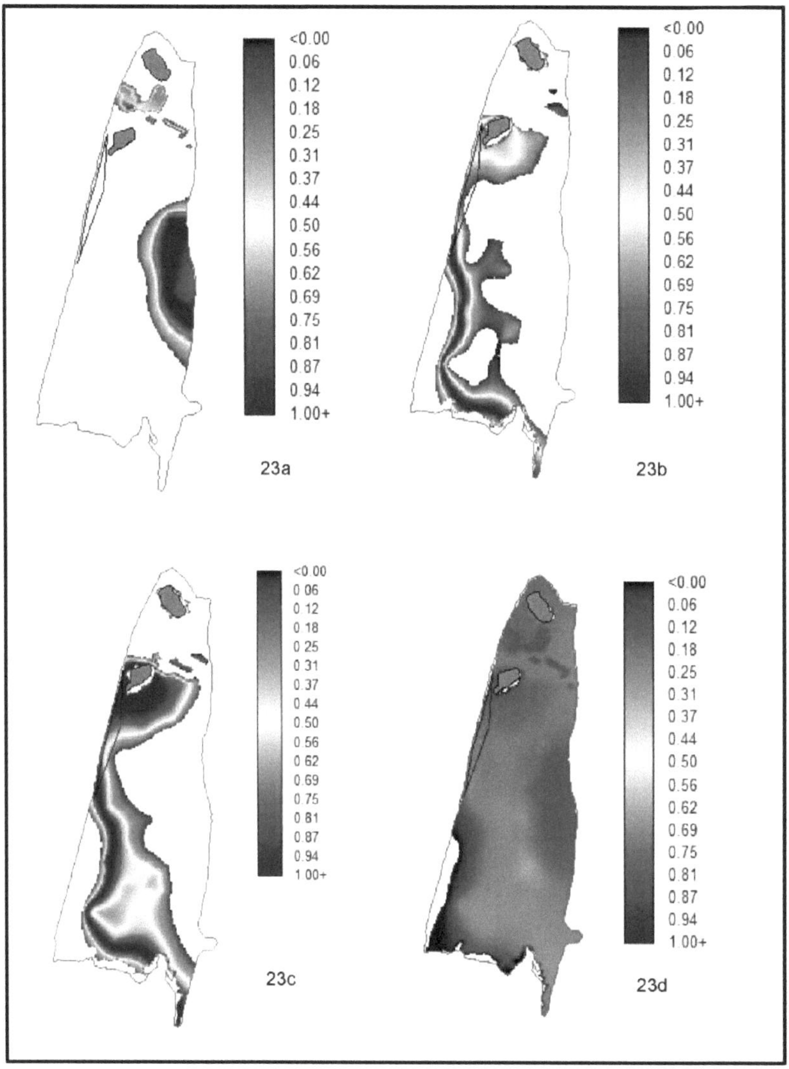


Figure 23 - Images generated through the environmental quality modeling process for *M. americana*. Figure 23a - Flood period. Figure 23b - Drought period. Figure 23c - Period with water levels comprised in the average of the data series. Figure 23d - Average BT suitability.

### 9.4.4 Validation of the AHI for *Caiman latirostris*

When analyzing a period with water levels comprised in the average of the data series (Figure 24c), it is possible to see in the species' distribution area that the environment is mostly favourable, with suitability equal to 1 (blue color), in terms of the height of the water level established in its AHI. This is not the case in extreme situations such as the periods of flooding (Figure 24a), with suitability between 0.44 and 0.06 (shades of orange and burgundy) and drought (Figure 24b), with a small region of the habitat area with suitability equal to 1. This helps corroborate reports that long periods of drought or flooding are detrimental to the species in the BT, due to the reduction in the availability of favorable habitat.

During the dry season, favorable sites for the yellow-bellied caiman are found on the banks of the Nicola and Jacaré lagoons, as observed in the field by Melo (1990), and in the southwestern portion of the marsh, a topographically lower region. The image representing the average water level (Figure 24d) was average, varying between 0.45 and 0.60 (shades of lighter orange and light green), demonstrating satisfactory suitability in some regions.

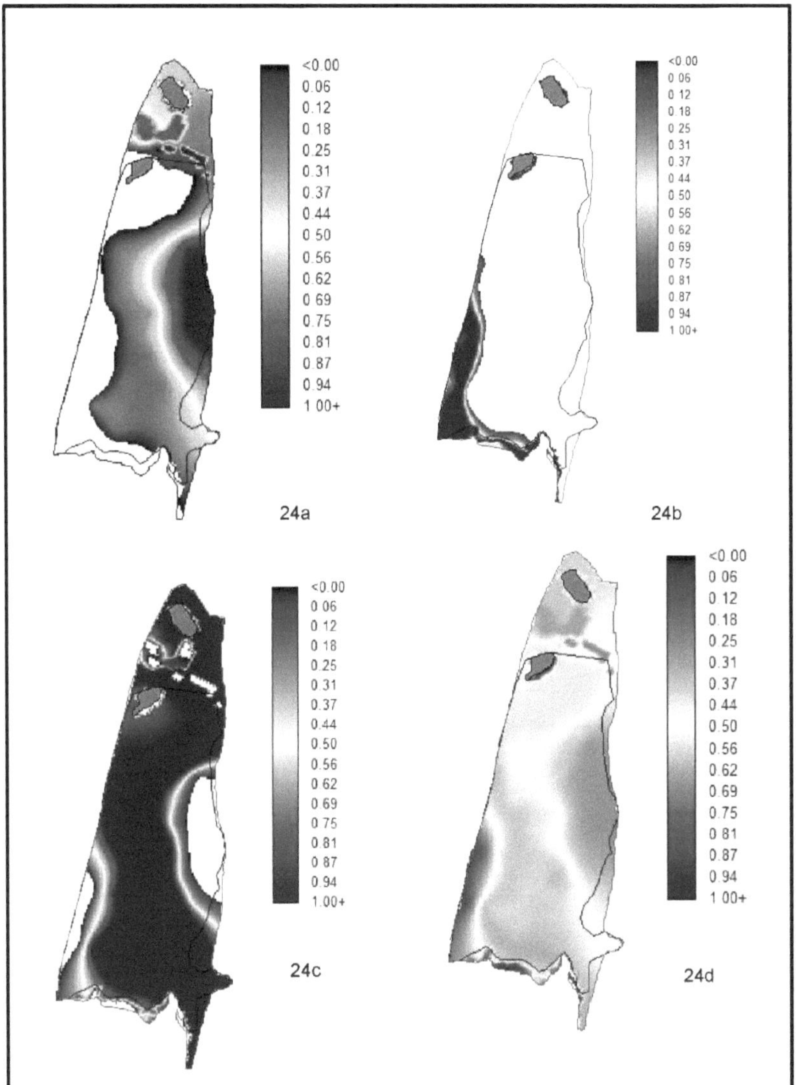


Figure 24 - Images generated through the environmental quality modeling process for *C. latirostris*. Figure 24a - Flood period. Figure 24b - Drought period. Figure 24c - Period with water levels comprised in the average of the data series. Figure 24d - Average BT suitability.

## 9.4.5 Validation of the AHI for *Trachemys dorbigni*

Due to the fact that the environmental suitability model is prepared to work with images with a spatial resolution of 101 x 101 m pixels, it was not possible to reproduce the 94 m of terrestrial habitat to be preserved around the bodies of water to guarantee the protection and integrity of the chelonian nests at ESEC Taim.

In the image corresponding to a period of flooding (Figure 25a), the regions around the Jacaré and Nicola lagoons and the inner part of the canal are characterized as being unfavorable for nest building. On the other hand, the outer part of the canal (between the canal and the highway), which is characterized by being a higher area, would guarantee the safety of the nests against flooding processes, even in times of flooding in the marsh.

During periods of drought (Figure 25b), some regions located around the lagoons would be suitable for maintaining nests, as well as almost the entire area around the canal, both inside and outside. As events of scarcity are seasonal, these regions would not be ideal for nesting, as these conditions must be maintained for at least 3 months, the period during which the eggs incubate and hatch. Periods of drought are characterized by a lack of food, which could lead to an increase in predation on nests located in these regions.

Under conditions with water levels at the average of the data series (Figure 25c), the most suitable regions for building nests are the sites in the outer part of the canal (between the canal and the highway), and the inner portion located further north in the canal. In this region, it is more likely that the nests will be protected from flooding during the egg incubation period between October and March.

The image depicting an average situation of the observed conditions (Figure 25d) shows that the most suitable place to guarantee the safety of chelonian nests in the Banhado do Taim region from flooding during the egg incubation period would be around the channel parallel to the BR-471 highway. Possibly this result is a consequence of the topographic gradient created by the side slope resulting from the construction of the highway.

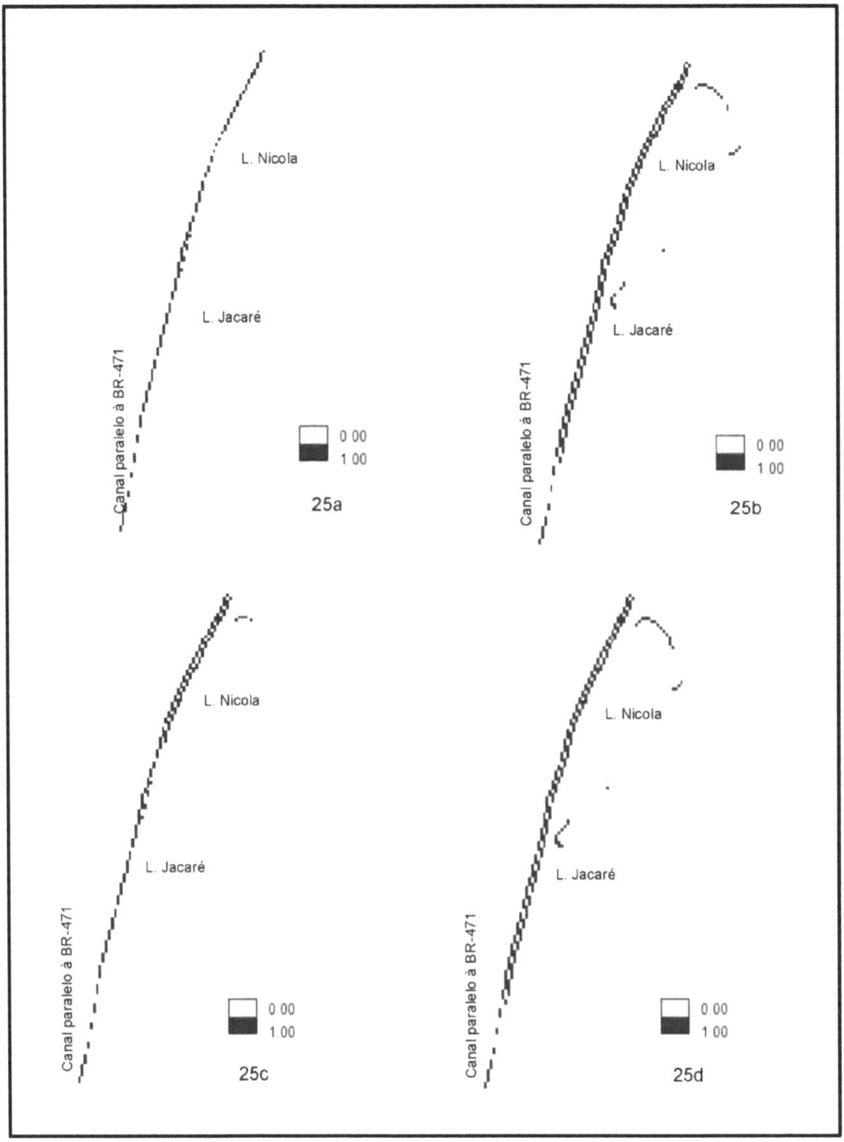


Figura 25 - Images generated through the environmental quality modeling process for P. chihi. Figure 25a - Flood period. Figure 25b - Drought period. Figure 25c - Period with water levels within the average of the data series. Figure 25d - Average BT suitability.

Considering the reproductive period, which extends from October to February, and the historical series with 44 years of data, a graph was drawn up (Figure 26) showing the percentage of area around the water bodies of interest (channel parallel to the BR-471 highway, Nicola and Jacaré lagoons) for *T. dorbigni* nesting in the Banhado do Taim region. As can be seen in this figure, the channel parallel to the BR-471 highway is the region that would retain the greatest number of nests under any hydrological condition. During periods of flooding, the lagoons are unable to accommodate the nests of this species, while during droughts, favorable conditions are created in the surrounding area.

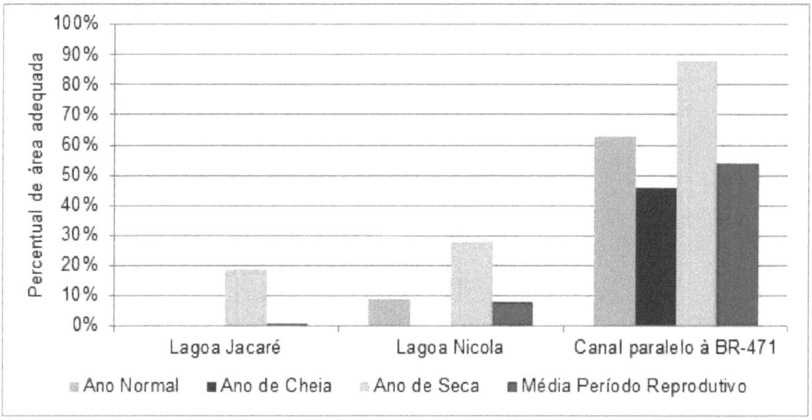


Figura 26 - Percentage of area around water bodies suitable for guaranteeing the integrity of *T. dorbigni* nests in the Banhado do Taim region during periods with water levels comprising the average of the data series, periods of flooding, periods of drought and the average of the reproductive period, corresponding to October and February of the historical series.

### 9.4.6 Comprehensive analysis of the AHI validation process

Through the validation process, it can be seen that the modeling of suitable habitats through the use of IAHs proved to be appropriate for Banhado do Taim. The most satisfactory images were those depicting periods with water levels within the average of the data series. The majority of the images depicting extreme periods such as drought and flood were not suitable for the indicators.

However, it should be emphasized that the species of this ecosystem are adapted to natural fluctuations in water levels, the big question is to determine how long such situations of water stress can be endured without compromising the populations and their capacity for resilience.

As for the average suitability images of the BT, it should be borne in mind that there are uncertainties associated with the topography of the bottom of the Banhado, which may have affected the representation of suitability for the indicators. In addition, it was not possible to determine habitat areas within the Banhado for some of the indicators representing birdlife, due to the difficulty of observing and accessing these sites.

An analysis of the suitability images shows that there are sites suitable for the occurrence of the species depending on the height of the water surface, but these were not determined as habitat areas. This may be related to other environmental factors not considered in this study, such as soil type, water flow velocity, food availability, predator occurrence and shelter availability (KREBS, 2001).

It is also likely that the individuals use different locations as foraging areas depending on the seasonality of the hydrological conditions of the environment. This situation could be observed during a field visit in April 2014, when the rulers located in the BT marked 0.79 m above the water level, and the black curlews were seen foraging in greater abundance on the banks of the Mangueira Lagoon.

## 9.5 Restructuring the IAHs and revalidation

In order to improve the environmental quality modeling process, the AHI was readjusted for the *S. californicus* indicator, whose suitability images did not show the expected result.

## 9.5.1 Restructuring the IAHs for *Scirpus californicus*

The first attempt to develop the AHI for *S. californicus* did not satisfy the reality observed in the marshland in terms of the region where the species occurs (shown in figure 21). To this end, the index was adjusted by checking the height of the average water table in the region mapped for *S. californicus,* which resulted in 1.00m. Thus, the new AHI was also based on the information on the limit water level for emergent macrophytes from Motta-Marques, Irgang and Giovannini (1997), resulting in a more accurate AHI than that observed in the BT (Figure 27).

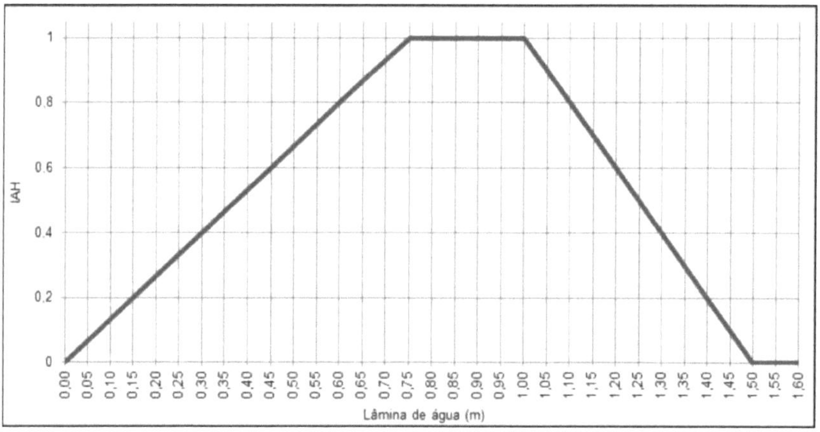


Figure 27 - Restructuring of the Habitat Suitability Index for the species *Scirpus californicus* for Banhado do Taim.

The index shows the tolerance of the minimum water level (between 0.10 m), passing through the optimum water level (1.00 m) that would guarantee the survival of the species, the effectiveness of its chemical processes such as photosynthesis, reproduction and interactions, up to its maximum limit. After a water depth of 1.40 m, which corresponds to around 70% of the size of the species' individuals, the environment would become critical and these processes could be compromised.

## 9.5.2 Revalidation of the IAHs for *Scirpus californicus*

When analyzing a period with water levels comprised in the average of the data series in the water levels of Banhado do Taim (Figure 28c), it is possible to recognize in the area of distribution of the species that the environment is mostly favorable for its occurrence depending on the height of the water level, with suitability between 1 and 0.5 (blue, dark green and yellow), with a small patch of low suitability, between 0.3 and 0 (gradient between orange and burgundy), which shows that the change in the index through its restructuring was adequate.

In extreme situations such as floods (Figure 28a) and droughts (Figure 28b), there is a representation of unfavorable environments for the species, with suitability varying between 0.3 and 0 (gradient between orange and burgundy), with small favorable areas of suitability between 0.5 and 1 (yellow, dark green and blue).

The image representing the average water clarity (Figure 28d) was not satisfactory for the species *S. californicus*, with suitability between 0.26 and 0.46 (orange color), but it was more favorable than the average image obtained for the first AHI developed for the species (Figure 21b).

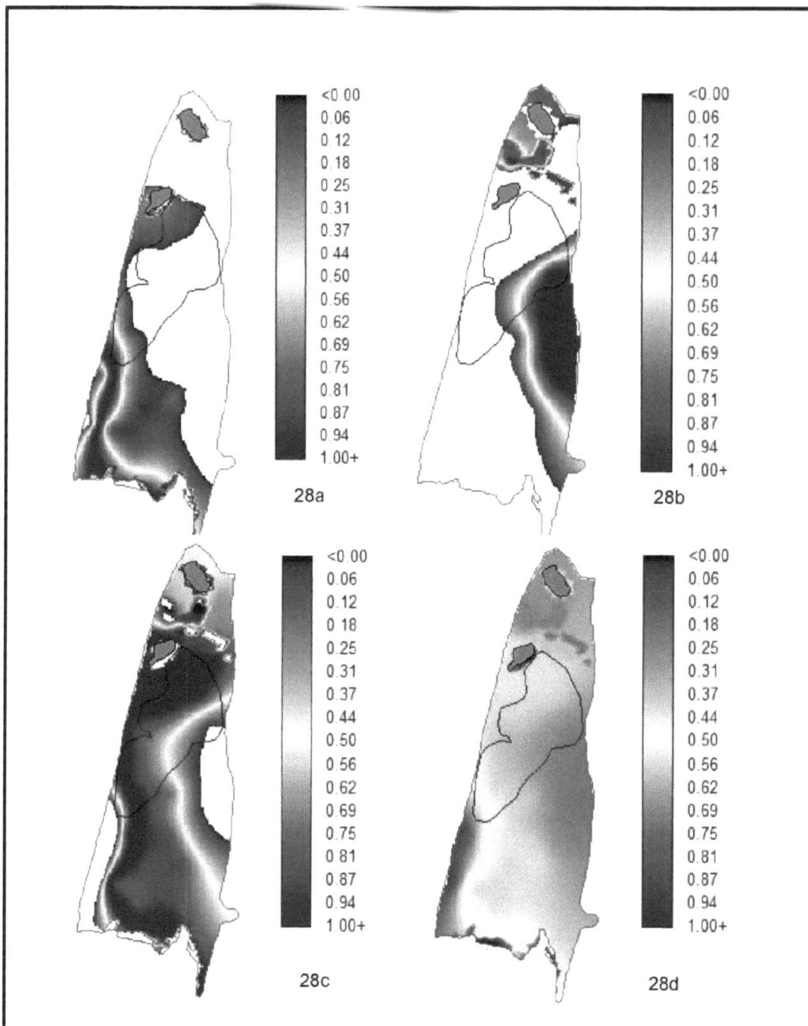


Figure 28 - Images generated through the environmental quality modeling process for *S. californicus*. Figure 28a - Flood period. Figure 28b - Drought period. Figure 28c - Period with water levels comprised in the average of the data series. Figure 28d - Average BT suitability.

## 9.6 Permanence curves

Permanence curves were drawn up for the indicators related to the height of the water level, information used to obtain water level values.

The criteria used to draw up the permanence curves (Figure 29) for each indicator are shown in Table 7.

Table 7 - Specification of the criteria used to draw up the permanence curves

| Indicator | Hydrodynamic model cells | Minimum water level (m) | Maximum water level (m) | Average water level (m) |
|---|---|---|---|---|
| *S. californicus* | 13, 17, 18, 21,23 e 25 | 02,31 | 03,42 | 02,77 |
| *P. chihi* | 25 | 01,84 | 03,10 | 02,74 |
| *M. americana* | 25 | 01,75 | 03,20 | 02,90 |
| *C. latirostris* | 03,04, 05, 08, 09, 10, 13, 14, 17, 18, 21, 22, 25 e 26 | 01,75 | 04,50 | 03,10 |

There was a great deal of variation in the minimum and maximum water level values found, which is related to the tolerance limits of each indicator in relation to the height of the water level, and also to its habitat area. Each habitat area covers different cells of the hydrodynamic model, which contains information on the water level of Banhado do Taim.

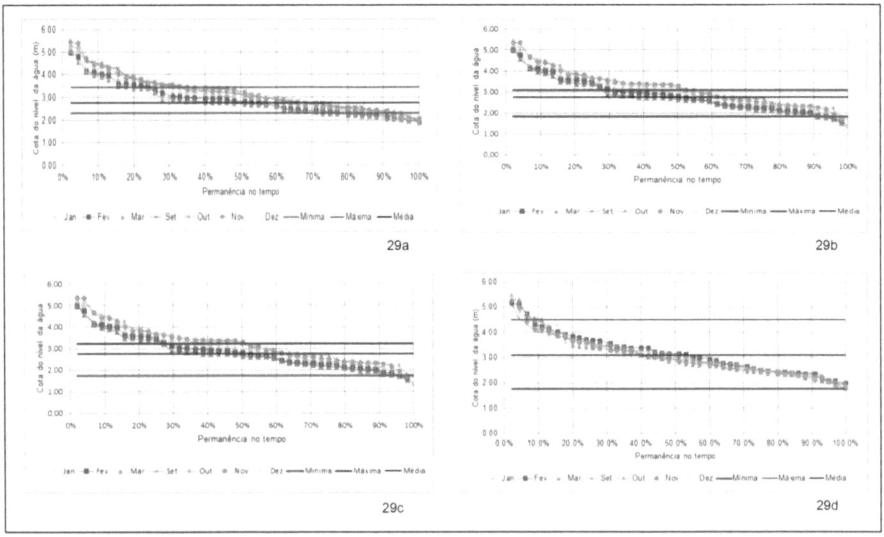

Figure 29 - Permanence curves developed for the indicators with an elaborated AHI as a function of water level. Figure 29a - *S. californicus* ; Figure 29b - *P. chihi* ; Figure 29c - *M. americana* ; Figure 29d - *C. latirostris*.

The permanence curves assign a degree of probability of the water level being equaled or exceeded. By analyzing the permanence curves, it is possible to consider that the maintenance of the minimum and maximum water levels is observed for a large part of the period analyzed, occurring at least 30 and 90% of the time. This shows that it would be possible to maintain the values obtained and that they could be included in the BT's water resources management process.

The months of February and March were more likely to present low water levels and exceed the values established as the minimum level. The months of October, November and December are critical for maintaining the maximum levels.

## 9.7 Proposed minimum and maximum water levels for Banhado do Taim based on bioindicators

The development of IAHs contributes to water resource management processes, serving as a subsidy in decisions regarding water withdrawals, and in resolving environmental and social issues regarding land use. It also helps to establish environmental flows and ecological levels, where a water regime is proposed to be sustained in order to conserve ecosystems and their benefits.

The organisms determined as indicators for Banhado do Taim are representatives of various groups, such as local vegetation, avifauna, herpetofauna and mastofauna. In Tassi's research (2008), AIs were developed for the bioindicators: *S. giganteus*, the swordtail (*Z. bonariensis)*, the black-necked swan (*C. melancoryphus*) and the capybara (*H. hydrochaeris)* (Figure 30). Combining these indices with those developed in this study: *S. californicus*, *P. chihi*, *M. americana*, *C. latirostris* and *T. dorbigni*, there are a total of nine suitability indices developed for Banhado do Taim. Of these, eight were generated according to the height of the water level, and were used for the analysis of the water level limits of the BT.

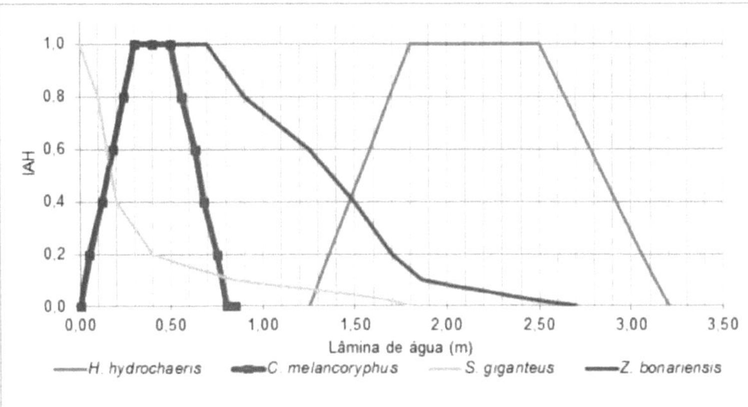


Figure 30 - Habitat suitability indices developed for the indicator species *Z. bonariensis*, *S. giganteus*, *C. melancoryphus* and *H. hydrochaeris*, as a function of water depth, for the Banhado do Taim wetland region. Adapted from Tassi (2008).

Figure 31 shows a summary of the water levels which, according to the defined AIs, can be considered for management purposes in Banhado do Taim. This figure includes both the values defined in this study and those established in the study by Tassi (2008). Figure 32 shows the relationship between the BT hydroperiod, determined by Tassi (2008), and the minimum and maximum levels for each bioindicator.

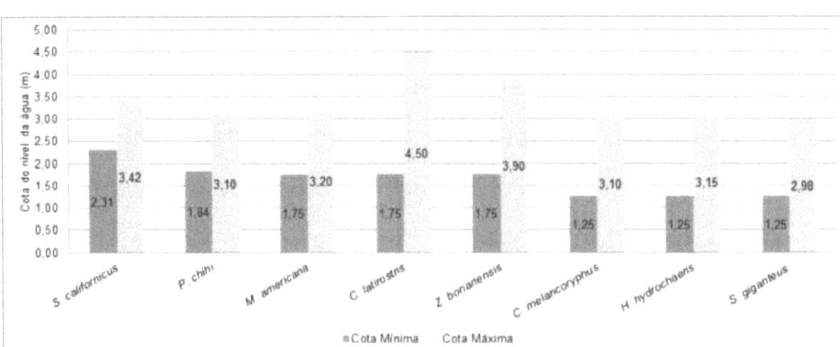


Figura 31 - Minimum and maximum water level limits for each indicator species used in the development of AIs as a function of water level height for Banhado do Taim.

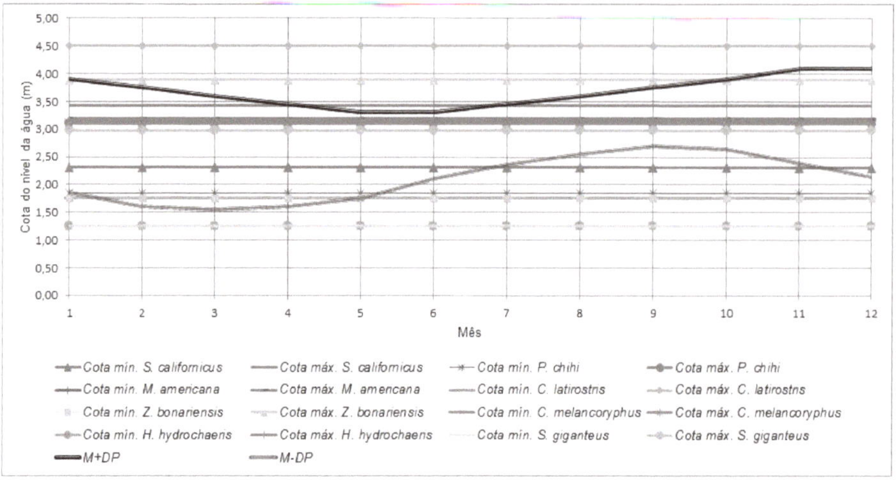


Figura 32 - Minimum and maximum elevation limits established for the indicator species, together with the seasonal variability of the Banhado do Taim hydroperiod. M - Mean; SD - Standard Deviation

By analyzing the data, it was possible to identify that the species that tolerate lower values of minimum water level (1.25 m) are the black-necked swan (*C. melancoryphus*), the capybara (*H. hydrochaeris)* and the macrophyte *S. giganteus*. However, this situation would no longer be satisfactory for the macrophyte *Z. bonariensis*, the yellow-bellied caiman (*C. latirostris*) and the migratory dry-headed bird (*M. americana).* The establishment of a minimum water level equal to 1.75 m would provide a favorable environment for all the indicator species, in addition to sustaining sufficiently suitable conditions for *M. americana,* which moves to the region in search of food between spring and summer. The value adopted ratifies the information on the minimum water level to be maintained in Banhado do Taim determined by Tassi (2008).

By comparing the value of the minimum water level obtained with the BT hydroperiod, it is possible to see that this limit is more likely to occur or be exceeded in the months of January to May. As already mentioned, this coincides with the period of greatest evapotranspiration, greatest water withdrawals for irrigation, and the reproductive phase of some species.

With regard to the maximum tolerable water level for the indicators, it can be seen that *C. latirostris* and *Z. bonariensis* stand out for their wide tolerance limits, but maintaining such high levels would be detrimental to the other species. In view of this, it is proposed that the lowest value accepted as the maximum tolerance be established as the maximum height, which is equivalent to 3.10 m, and corresponds to the birds *P. chihi* and *C. melancoryphus*.

Certainly, a maximum level of less than 3.10 m would ensure a favorable environment for the other bioindicators, as it does not exceed any of the tolerance values for flooding, and is above the average level, corresponding to 2.90 m. In addition, the limit value of the maximum level is incorporated into the BT's annual hydroperiod, not exceeding the (M+DP) delimitation curve, and is more likely to be exceeded in the months of January, September, October, November and December.

This is related to the flooding periods of Banhado do Taim, which occur approximately 3 months after the peak of rainfall (MOTTA-MARQUES *et al.*, 2013). Although rainfall is well distributed throughout the year, according to the historical series, the highest volumes of rain generally occur between the months of June and September.

The value of the maximum water level proposed for the BT is above the value established by Tassi (2008), defined as 2.98m. This shows that the more research is carried out, and the greater the number of IAHs developed, the more refined the process of defining the minimum and maximum water levels to be maintained in Banhado do Taim will be.

The water levels defined as minimum and maximum (1.75m and 3.10m) should be maintained as a form of control, especially in situations of extreme or atypical phenomena, as it is essential to respect the natural oscillations in the water levels of the wetland.

By analyzing the series of water levels in the BT for a period of 44 years

(1960 to 2003 - Figure 33), it is possible to identify the natural oscillations in the water levels inside the marsh over time. On average, the levels vary in 3-year intervals between values above or below the average level. However, it is possible to delineate long periods of flood and drought, represented by a period of 11 (1977 and 1988) and 9 years (1989 and 1998), respectively.

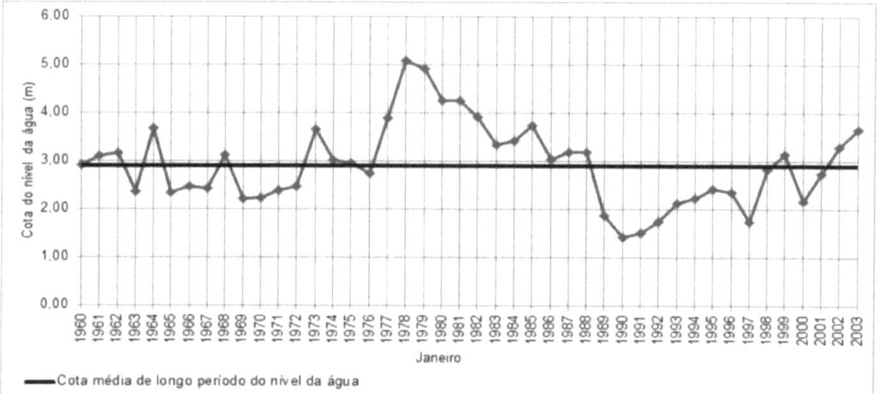


Figure 33 - Water level series of Banhado do Taim - period January 1960 to January 2003.

Through this analysis, it is possible to infer that the indicator species would be able to withstand long periods of water stress. However, there are no specific studies on the effects of these cycles on the organisms, how they affect their development, reproductive rate, survival or competition.

Under critical or adverse situations, individuals representing avifauna, mastofauna and herpetofauna can move in search of favorable conditions. For plants, the consequences can occur through changes in morphophysiological behavior (SANTOS; CARLESSO, 1998). However, more studies are needed on the BT's flooding pulses in order to establish the period of time in which the community of organisms would be able to withstand situations close to or beyond their tolerance limits.

## io CONCLUSIONS AND FINAL CONSIDERATIONS

Through this research, it was possible to develop five new IAHs for Banhado do Taim, namely: the reed bunting (*Scirpus californicus*), the black-capped sandpiper (*Plegadis chihi)* and loggerhead (*Mycteria americana)* birds, and the reptiles yellow-bellied *caiman* (*Caiman latirostris*) and tiger turtle (*Trachemys dorbigni*).

The definition of the indicator species was based mainly on their observation in the field, and especially on the availability of hydrological information: the height of the water level needed to guarantee their survival and the area needed to be kept free from flooding during the reproductive period. In order to evaluate situations that guarantee the performance of chemical processes, the availability of food and the reproductive success of these species.

It should be noted that there is a scarcity of studies on the environmental conditions favorable to the survival of species, specifically in terms of hydrology. It would be extremely important to encourage more research into the eco-hydrology of Banhado do Taim, since it has a great wealth of species, especially birds. Among the groups of birds, the Anatidae Family stands out for its need for a very restricted water level, as they feed by filtering the water, inserting their necks below the surface.

In the process of validating the suitable habitat modeling process, the AIs generally showed good suitability in relation to the areas determined as habitat, as reflected in the images analyzed. Even so, it is proposed to improve the representation of the MNT, in particular the region of the canal that runs parallel to the BR-471 highway, which proved to be a very important region for most of the indicator species, which could reflect a better suitability for the AHI.

The final objective of the study was to establish the minimum and maximum water levels to be maintained in Banhado do Taim in situations of water stress, a process strictly related to the suitability of the environment for the indicator species. It is essential to preserve the hydroperiod and hydrodynamics of the site, but in unfavorable situations, the level limits determined in this study can be used as a form of sustainable management.

Determining the minimum level of 1.75 m and the maximum level of 3.10 m can help limit the withdrawal of water from the region for irrigating crops during dry periods when there is less water available, respecting the water level heights to be maintained in the BT to ensure a favorable environment for the wetland community. However, in situations where the wetland is flooded and is expected to remain flooded for a long period, it would be possible to increase the use of this resource for irrigation in situations that exceed the limits of the maximum level, as also proposed by Tassi (2008).

Even so, for this management process to take place properly, it would be essential to identify the long periods of water stress, which boost the ecosystem's resilience, since alterations to this environment could change the configuration and predominance of species in the BT. It would also be advisable to monitor water withdrawals at the hydrographic basin level, as the water level in Banhado do Taim also depends on the levels of Lagoa Mangueira.

Finally, the development of new AIs, as well as the generation of composite AIs, could provide a more accurate picture of the various factors that influence and determine the suitability of Banhado do Taim. This will contribute more eco-hydrological information, helping more precisely in the analysis of adequate water levels to be maintained. This information will also support water resource management processes through sustainable management, which will guarantee the conservation and preservation of the ecosystem and its functions, as well as the multiple uses of this essential resource.

## 10. REFERENCES

ANA. National Water Agency. **Caring for water: solutions for improving the quality of water resources**. Brasilia: National Water Agency/United Nations Environment Program, 2011. 152 p.

ARMITAGE, R. A.; FONG, P. Gastropod colonization of a created coastal wetland: potential influences of habitat suitability and dispersal ability. **Restoration Ecology**. Malden, v. 12, n. 3, p. 391- 400. September, 2004.

BABBITT, K. J. The relative importance of wetland size and hydroperiod for amphibians in southern New Hampshire, USA. **Wetlands Ecology and Management**. Dordrecht, v. 13, p. 269-279. 2005.

BAGER, A. **Aspects of the biology and ecology of the water tiger turtle *Trachemys dorbignyi* (Testudines - Emydidae) in the extreme south of the state of Rio Grande do Sul - Brazil**. 2003. 100 f. Thesis (Doctorate in Ecology) - Federal University of Rio Grande do Sul. Porto Alegre, 2003.

BAGER, A. **Aspects of the reproductive dynamics of *Phrynops hilarii* (Duméril e Bibron, 1835) (Testudines - Chelidae) in southern Rio Grande do Sul**. 1997. 120 f. Dissertation (Master's Degree in Ecology) - Federal University of Rio Grande do Sul, Porto Alegre, 1997.

BAGER, A.; ROSADO, J. L. O. Estimation of core terrestrial habitats for freshwater turtles in southern brazil based on nesting areas. **Journal of Herpetology**. Athens, v. 44, n. 4, p. 658-662. 2010.

BAKER, R. J.; SHONEWALD-COX, C. M. Management strategies for improving population viability. In: WILCOX, B. A.; BRUSSARD, P. F.;

MARCOT, B. G (Ed.). **The Management of Viable Populations:** theory, applications and case studies. Center for Conservation Biology, Stanford University, Palo Alto, p. 73-87.1986.

BARBIER, E. B.; ACREMAN, M. C; KNOWLER, D. **Economic valuation of wetlands: A guide for policy makers and planners**. Switzerland: Ramsar

Convention Bureau, 1997. 116 p.

BENCKE *et al.* Revised and updated list of the birds of Rio Grande do Sul, Brazil. **Iheringia - Zoology Series**. Porto Alegre, v. 100, n. 4, p. 519-556, 2010.

BOBBINK, R. *et al.* Wetlands: functioning, biodiversity conservation, and restoration. **Ecological Studies**. Berlin, v. 191, 2006. 315 p.

BORGES-MARTINS, M. *et al.* Reptiles. In: BECKER, F.G.; RAMOS, R.A.;

MOURA, L.A. (Orgs.). **Biodiversity: Lagoa do Casamento and Butiazais de Tapes Regions, Coastal Plain of Rio Grande do Sul**. Brasilia: Ministry of the Environment, 2007. Chap. 21, p. 292-315.

BORKHATARIA *et al.* Temporal variation in local wetland hydrology influences post dispersal survival of juvenile wood storks (*Mycteria americana*). **The Auk**. Lawrence, v. 129, n. 3, p. 517-528. 2012.

BRAZIL. **Institutes the National Water Resources Policy** (1997). Law

Federal Law n. 9433, of January 8, 1997. Published in the Federal Official Gazette on January 9, 1997, section 1, p. 470, v. 135, n. 6.

BRAZIL. **National System of Conservation Units** (2000). Federal Law n. 9985. Brasilia, July 18, 2000. Available at:< http://www.planalto.gov.br/ccivil_03/leis/l9985.htm>. Accessed on: March 4, 2013.

BRATTEBO, S. K. *et al.* Development of Periphyton, Benthic Macroinvertebrate, and Macrophyte Habitat Suitability Indices for Assessment of Reservoir Quality and Potential Production on the Pend Oreille River. In: Waterpower XVI Copyright PennWell Corporations. Washington, 2009.

**Annals of Congress**.14 p. 2009.

BRAVO, J. M. *et al.* Evaluation of temperature and precipitation anomaly projections in future climate scenarios in the region of the Taim hydrological system, RS. In: XIX Brazilian Symposium on Water Resources. **Proceedings**. Maceió, 2011.

BROTONS *et al*. Presence-absence versus presence-only modelling methods for predicting bird habitat suitability. **Ecography**. Copenhagen, v. 27, p. 437-448, 2004.

BROWN, J. R.; HARTWICK, E. B. A habitat suitability index model for suspended tray culture of the Pacific oyster, *Crassostrea gigas* Thunberg. **Aquaculture Research**. Oxford, n. 19, v. 2, p. 109-126. 1988.

BUNN, S. E.; ARTHINGTON, A. H. Basic principles and ecological consequences of altered flow regimes for aquatic biodiversity.

**Environmental Management**. New York, n. 30, p. 492-507. 2002.

BURGER, M. I. **Situation and priority actions for the conservation of coastal wetlands**. Porto Alegre: RS Zoobotanical Foundation, 2000. 60 p.

BURGER, M. I.; RAMOS, R. A. Important areas for conservation in the Coastal Plain of Rio Grande do Sul. In: BECKER, F. G.; RAMOS, R. A.; MOURA L. de A. (Org.). **Biodiversity: Lagoa do Casamento and Butiazais de Tapes Regions, Coastal Plain of Rio Grande do Sul**.

Ministry of the Environment, Brasilia, p. 46-56. 2007.

CABRERA, M. R. **Las tortugas continentales de Sudamérica Austral**.

Córdoba: BR Copies, 1998. 108 p.

CARREIRA, S.; MENEGHEL, M.; ACHAVAL, F. **Reptiles of Uruguay**.

Montevideo: Edición División Relaciones y Actividades Culturales, Faculdad de Ciencias, Universidad de la Repûblica, 2005. 639 p.

CARVALHO, A. B. P.; OZORIO, C. P. Assessment of the wetlands of Rio Grande do Sul, Brazil. **Journal of Environmental Sciences**. Canoas, v. 1, n. 2, p. 83-95. 2007.

CHOMENKO, L. Water resources and wetlands: environments to be preserved. In: ZAKRZEVSKI, S. B. (Org.). **Conservation and sustainable use of water: multiple perspectives**. Erechim: EdiFapes, RS. 2007. 138 p.

CISRERP. Committee on Independent Scientific Review of Everglades Restoration Progress. **Progress Toward Restoring the Everglades: The Fifth Biennial Review, 2014**. Washington: National Research Council of the National Academies. The National Academies Press, 2014. 240 p.

CORDAZZO, C. V.; SEELIGER, U. **Guia ilustrado da vegetaçâo costeira do extremo sul do Brasil**. Rio Grande: FURG, 2. ed., 1988. 275p.

CORREA, F. *et al.* Pisces, Gymnotiformes, Hypopomidae, *Brachyhypopomus gauderio* Giora and Malabarba, 2009: New species record at Taim Ecological Reserve, southern Brazil. **Check List**. v. 7, n. 1, p. 19-20. 2011. Available at: < http://www.checklist.org.br/archive?vol=7&num=1>. Accessed on: September 26, 2014.

COULTER, M. C. *et al.* Wood Stork (*Mycteria Americana*). In: The American Ornithologists' Union (Ed.). **The birds of North America**. Ithaca: Cornell Lab of Ornithology; Retrieved from the Birds of North America. n. 409, 1999. p. 128.

COULTER, M. C.; BRYAN, Jr. A. L. Foraging ecology of Wood Storks (*Mycteria americana*) in east-central Georgia. I. Characteristics of Foraging Sites. **Colonial Waterbirds**, De Leon Springs, n. 112, p. 237-243. 1993.

DALE, V. H.; BEYELER, S. C. Challenge in the development and use of ecological indicators. **Ecological Indicators**. *Athens,* v. 1, p. 3-10. 2001.

DARK-SMILEY, D.; KEINATH, D. A. **Species assessment for white-faced-ibis (*Plegadis chihi*) in Wyoming**. Wyoming: United States Department of the Interior Bureau of Land Management Wyoming State Office Cheyenne. 2003. 59 p.

DEPKIN, F. C.; COULTER, M. C.; BRYAN JR, A. L. Food of nestling wood stork in east-central Georgia. **Colonial Waterbirds**. De Leon Springs, n. 15,

p. 219-225. 1992.

DONAVAN, T. M. *et al.* Quantifying home range habitat requirements for bobcats (*Lynx rufus*) in Vermont, USA. **Biological Conservation**. Essex, n. 144, p. 2799- 2809. 2011.

DUBERSTEIN, C. A. *et al.* **Development of a Habitat Suitability Index Model for the Sage Sparrow on the Hanford Site**. Washington: U.S. Department of Energy under Contract/Pacific Northwest National Laboratory Richland, 2008. 25 p.

ESTEVES, F. A. **Fundamentos de Limnologia**. Rio de Janeiro: Interciência, 3. ed., 2011. 826 p.

FERNANDES, I. M. *et al.* **Biodiversity in the Poconé Pantanal**. Cuiabà:

Attema Design Editorial/Centro de Pesquisa do Pantanal, 2010. 196 p.

FERREIRA, T. F. **Ecology and applicability of methods for evaluating the primary production of *Zizaniopsis bonariensis*: an emerging aquatic macrophyte, in the Taim Hydrological System, RS**. 2005. 135 f. Dissertation (Master's in Ecology) - Federal University of Rio Grande do Sul. Porto Alegre, 135 f, 2005.

FETT, M. S. **Brazilian Technical Answers Service**. Porto Alegre: SENAI - RS, September 29, 2005. Available at:<http://www.sbrt.ibict.br>. Accessed on: August 12, 2014.

FLORIDA FISH AND WILDLIFE CONSERVATION COMMISSION. **Plants for lakefront revegetation**. Florida: Division of Habitat and Species Conservation. 2002. 25 p.

FRAGOSO-JÙNIOR, C. R. *et al.* Simulation of Eutrophication in Shallow Lakes

II: Taim System (RS). **Brazilian Journal of Water Resources**. Porto Alegre, v. 12, n. 4, p. 37-48, Oct/Dec, 2007.

GARCIA, A. M. *et al.* Preliminary examination of food web structure of Nicola Lake (Taim Hydrological System, southern Brazil) using dual C and N stable isotope analyses. ***Neotropical Ichthyology***. Porto Alegre, RS, v. 4, n. 2, p. 279-284, 2006.

GAWLIK, D. E. *et al.* **A summary of recent studies showing how prey availability affects wading bird habitat selection, physiology, and production**. Homestead: South Florida Natural Resources Center, Everglades National Park. 2010.

GAYER, S. M. P.; KRAUSE, L.; GOMES, N. Preliminary list of the amphibians of the Taim Ecological Station. **Brazilian Journal of Zoology**. Sao Paulo, v. 5, n. 3, p. 419- 425. 1988.

GIANNINI, T. C. *et al.* Review article/Current challenges of predictive modeling of species distribution. **Rodriguésia**. Rio de Janeiro, v. 3, n. 63, p. 733-749. 2012.

GIOVANNINI, S. G. T. **Estudo da Variabilidade espacial da vegetaçâo no banhado do Taim**. 2004. 267 f. Thesis (Doctorate in Water Resources and Environmental Sanitation) - Federal University of Rio Grande do Sul. Porto Alegre. 2004.

GOMES, N.; KRAUSE, L. Preliminary list of reptiles from the Taim Ecological Station, Rio Grande do Sul. **Brazilian Journal of Zoology**. Sâo Paulo, v. 1, n. 1, p. 71-77. 1982.

GOPAL, B.; JUNK, W. J. Biodiversity in wetlands: an introduction. In: GOPAL, B.; JUNK, W. J.; DAVIS, J. A. (Eds.). **Biodiversity in Wetlands: Assessment, Function and Conservation**. Leiden: Backhuys Publishers, v. 1, p. 1-10. 2000.

GRINNELL, J. Field tests of theories concerning distributional control. **The American Naturalist**. Chicago, v. 51, n. 602, p. 115-128, Feb 1917.

GUASSELI, A. L. **Vegetation dynamics in the Taim marshland, RS**. 2005.

157 f. Thesis (Doctorate in Water Resources and Environmental Sanitation) - Federal University of Rio Grande do Sul. Porto Alegre, 2005.

GUTZWILLER, K. J.; ANDERSON, S. H. Habitat suitability index models: marsh wren. U.S. Fish and Wildlife Service. **Biological Report**. *Washington,* June, 1987. 13 p.

HAHN, A. T. **Diet analysis of *Trachemys dorbigni* (Duméril e Bibron, 1835) in southern Rio Grande do Sul, Brazil (Testudines: Emydidae)**. 2005.

53 f. Dissertation (Master's Degree in Animal Biology) - Federal University of Rio Grande do Sul. Porto Alegre, 2005.

HIRZEL, A. H.; LAY, G. Habitat suitability modelling and niche theory.

**Journal of Applied Ecology**. Oxford, v. 45, p. 1372-1381. 2008.

HUTCHINSON, G. E. Concluding remarks. **Cold Spring Harbor Symposia on Quantitative Biology**. New York, n. 22, p. 415-427. 1957.

HYDRA AQUATIC INC. **Plant Catalog**. Albuquerque, 2006. 8 p. Available at: <http://www.hydraaquatic.com/catalog.htm>. Accessed on: 27 mer. 2014.

IBAMA. Brazilian Institute for the Environment and Renewable Natural Resources. **Wetlands**. Available at: <http//www.ibama.gov.br.> Accessed on: August 6, 2013. 2000.

IUCN. International Union for the Conservation of Nature and Natural

Resources.**Vision for water and nature - a world strategy for conservation and sustainable management of water resources in the 21st century**. Canada. 2000. 58 p.

JUNK, W. J. **Definition and Classification of Brazilian Wetlands**.

Available

at:<http://www.inau.org.br/classificacao_areas_umidas_completo.pdf>.
Accessed on: Aug. 27, 2013.

JUNK, W. J. Wetlands of Tropical South America. In: WHIGHAM, D.; HEJNY, S.; DYKYJOVA, D. (Eds.): **Wetlands of the world**. Dordrecht, p. 679-739. 1993.

JUNK, W. J. *et al.* **The Pantanal: Ecology, Biodiversity and Sustainable Management of a large Neotropical Seasonal Wetland**. Sofia: Pensoft Publishers, 2011. 857 p.

JUNK, W. J.; WANTZEN, K. M. The flood pulse concept: new aspects, approaches and applications - an update. In: WELCOMME, R.; PETR, T. (Eds.). **Proceeding of international Symposium on the Management of Large River for Fisheries**. Bangkok: RAP Publications/FAO, p. 117 - 149. 2004.

KISSMANN, K. G. **Weeds and Noxious Plants**. Volume I. Sao Paulo: Basf, 2. ed., 1997.825 p.

KRAUSE, L.; GOMES, N.; LEYSER, L. Observations on the nesting and development of *Chrysemys dorbigni* (Dumeril & Bibron, 1835) (Testudines, Emydinae) in the Taim Ecological Station, Rio Grande do Sul. **Brazilian Journal of Zoology**. Sao Paulo, v. 1, n. 1, p. 79-90. 1982.

KREBS, C. J. **Ecology**: **the Experimental Analysis of Distribution and Abundance. Addison Wesley Longman.** San Francisco: Benjamin Cummings, 5. ed., 2001. 695p.

KUSHLAN, J. A.; BILDSTEIN, K. L. White Ibis. In: POOLE, A.;

STETTENHEIM, P.; GILL, F. (Eds.). **The Birds of North America**.

Washington: Academy of Natural Sciences of Philadelphia and American Ornithologist's Union. n. 9, 1992.

LARSSON, M. A. **Landscape-level habitat suitability models for twelve wildlife species in southern Missouri**. St. Paul: U.S. Department of Agriculture/Forest Service/North Central. 2003. 51 p.

LEMA, T.; FERREIRA, M. T. S. Contribuiçao ao conhecimento dos Testudines do Rio Grande do Sul (Brasil) - Lista sistemàtica comentada (Reptilia). **Acta Biologica Leopoldensia**. Sao Leopoldo, v. 12, p. 125-164. 1990.

LO GALBO, A. *et al.* **Wood Stork Foraging Probability Index (STORKI v. 1.0) Ecological and Design Documentation** - Ecological and Design Documentation. 2012. 22 p. Available at:< http://www.cloudacus.com/simglades/docs/WoodStork Ecologicaldoc_20120601.pdf >. Accessed on:<September 21, 2014>.

LO GALBO, A. M. *et al.* Using hydrologic suitability for native Everglades slough vegetation to assess Everglades restoration scenarios. **Ecological Indicators**. *Athens,* n. 24, p. 294-304. 2013.

LOBO, J. M.; JIMÉNEZ-VALVERDE, A.; REAL, R. AUC: a misleading measure of the performance of predictive distribution models. **Global Ecology and Biogeography**. Oxford, v. 17, p. 145-151. 2008.

LOPES, I. F. **Genetic variability in populations of *Jabiru mycteria* (Lichtenstein, 1819) and *Mycteria americana* (Linneaus, 1758) (Aves, Ciconiidae): gene flow and phylogeography**. 2006. 106 f. Thesis (Doctorate in Genetics and Evolution) - Federal University of Sao Carlos. Sao Carlos, 106 f. 2006.

LOPEZ, R. D.; HEGGEM, D. T. **An ecological and habitat vulnerability assessment of the white river basin**. Las Vegas: U.S. Environmental Protection Agency National Exposure Research Laboratory/Environmental Sciences Division/ORD, 2001. 24 p.

LUZ, L. D. **Hydrology and habitat suitability assessment of a freshwater wetland**. 2002. 158 f. Dissertation (Doctor of Philosophy) - Cornell University, Ithaca, New York, USA. 2002.

MAHLER Jr, J. K. F.; KINDEL, A.; KINDEL, E. A. I. Commentary list of the bird species of the Taim ecological station, Rio Grande do Sul, Brazil. **Acta Biologica Leopoldensia**. Sao Leopoldo, v. 18, n. 1, jan./jun., p. 69103. 1996.

MARENGO, J. A.; SILVA DIAS, M. Global climate change and its impact on water resources. In: REBoUçAS, C. A.; BRAGA, B.;

TUNDISI, J. G. **Fresh waters in Brazil - ecological capital multiple uses, rational exploitation and conservation**. Sâo Paulo: IEA/USP, 3. ed., p. 63-109. 2006.

MCCARTNEY, M. *et al.* **Wetlands, agriculture and poverty reduction**. Sri Lanka: Colombo International Water Management Institute. IWMI Research Report 137. 39p. 2010.

MELO, M. T. Q. **Diet of *Caiman latirostris* (Daudin, 1802) (Crocodilia: Alligatoridae) in the Taim ecological station, RS**. 1990. 63 f. Dissertation (Master's Degree in Biosciences) - Pontifical Catholic University of Rio Grande do Sul. 63 p. 1990.

MITSCH, W. J.; GOSSELINK, J. G. **Wetlands. New York: Van Nostrand Reinhold Company Inc.**, 2nd ed. 1993. 722 p.

MMA. Ministry of the Environment; ICMBio. Chico Mendes Institute for Biodiversity Conservation; Directorate for the Creation and Management of Conservation Units Taim Ecological Station. **Executive summary of the proposal to expand the Taim Ecological Station**. September 2013. 25 p. Available at:< http://www.icmbio.gov.br/portal/images/stories/o- que-fazemos/consultas_publicas/ RESUMO_EXECUTIVO_Ampliacao_da_ESEC_do_Tai m.pdf>. Accessed on:September 01, 2014>.

MOLINA, F. B.; GOMES, N. Artificial incubation of eggs and hatching process in *Trachemys dorbigni* (Duméril e Bibron) (Reptilia, Testudines, Emydidae). **Brazilian Journal of Zoology**. Sâo Paulo, v. 15, p. 135-143. 1998.

MORA, J. W.; MAGER, J. N.; SPIELES, D. J. Habitat and Landscape Suitability as Indicators of Bird Abundance in Created and Restored Wetlands. International Scholarly Research Network. **ISRN Ecology**. 2011. 10 p. Available at:< http://www.hindawi.com>. Accessed on: May 22, 2014.

MORREALE, S. J.; GIBBONS, J. W. Habitat suitability index models: slider turtle. U. S. Fish and Wildlife Service. **Biological Report**. *Washington,* n. 82, 1986. 14 p.

MOTTA-MARQUES *et al.* The Taim Hydrological System - site 7. In

SEELIGER, U. *et al.* (Edit.). **The sites and the Brazilian program of long-term ecological research**. Belo Horizonte: Universidade Federal de Minas Gerais/Fundaçâo Universidade do Rio Grande, p. 125-141.2002.

MOTTA-MARQUES *et al.* The Taim hydrological system. In: LACERDA, L.

D. *et al.* **PELD - CNPQ: ten years of the Long-Term Ecological Research program in Brazil: findings, lessons and perspectives**. Recife: Federal University of Pernambuco. 2013. 446p.

MOTTA-MARQUES, D. M. L.; IRGANG, B.; GIOVANNINI, S. G. T. The importance of hydroperiod in water management in wetlands with multiple use - the case of the Taim Ecological Station. 1997, Vitória. **Anais**. XII Simpòsio de Recursos Hidricos. p. 1-7, 1997.

MOTTA-MARQUES, D.; VILLANUEVA, A. O. N. The hydrological regime of wetlands and their conservation. **Caderno de Pesquisa Biology Series**. Santa Cruz do Sul, n. 13, p. 63-79. 2001.

MOULTON, T. P.; MAGNUSSON, W. E.; MELO, M. T. Q. Growth of *Caiman latirostris* inhabiting a coastal environment at Ilha do Cardoso, SP, Brazil.

**Journal of Herpetology**. Athens, v. 33, p. 479-484. 1999.

MUSHET, D. M.; EULISS Jr., N. H.; STOCKWELL, C. A. Mapping Anuran Habitat Suitability to Estimate Effects of Grassland and Wetland Conservation

Programs. **Copeia**. Lawrrence, n. 2, p. 321-330. 2012.

NEMA. Environmental Education and Monitoring Center. **Sustainable Development Plan for the communities surrounding the Taim Ecological Station - RS**. Rio Grande: PROBIO/MMA. September, 2002. 85 p.

NETO, W. L. M.; PINTO, J. G.; CAMPOS, A. B. Norms and social conflicts in the management of wetlands in conservation units: the case of Serra da Areia Municipal Park in Aparecida de Goiânia, Goiàs. **Boletim Goiano de Geografia**. Goiàs, v. 31, n. 1, p. 97-112, jan./jun. 2011.

NILSSON, C. *et al.* Restoring riverine landscapes: the challenge of identifying priorities, reference states, and techniques. **Ecology and Society**, Atlanta, v. 12, n. 16. 2007.

NIMER, E. Clima. In: **Geography of Brazil**. Rio de Janeiro: IBGE Foundation, v. 2, p. 47-84. 1977.

O'NEIL, L. J. *et al.* A procedure to modify Habitat Suitability Index models.

**Wildlife Society Bulletin**. Bethesda, v. 16, n. 4, p. 33-36. 1988.

ODUM, E. P. **Fundamentals of Ecology**. Lisbon: Calouste Gulbenkian Foundation, 7. ed. 2004.

PANESSA, R.; PALO JR., H.; TRESCA, D. **Taim Ecological Station**.

Lloyds Bank cultural project. Sâo Paulo: Empresa das Artes, 1989. 44 p.

PAZ, A. R. **Analysis of the main factors involved in the hydrodynamic behavior of the Taim marsh (RS).** 2003. 176 f. (Master's Degree in Water Resources and Environmental Sanitation) - Federal University of Rio Grande do Sul, Porto Alegre. 2003.

PERONI, N.; HERNANDEZ, M. I. M. **Ecologia de Populaçoes e Comunidades**. Florianópolis: CCB/EAD/UFSC. 2011. 125 p.

PETERSON, A. T. Uses and requirements of ecological niche models and related distributional models. **Biodiversity Informatics**. Kansas, n. 3, p.

5972. 2006.

PINHEIRO, M. S.; SILVA, J. J. C. **Characterization of Birds Chosen to Evaluate Sustainability Indicators in the Southern Coastal Plain of Rio Grande do Sul**. Pelotas: Embrapa Clima Temperado, 2013. 28 p.

PINTO, L. F. S. *et al.* Characterization of vàrzea soils. In: **Manejo do solo e da água em áreas de vàrzea**. Pelotas: Embrapa Clima Temperado, Documentos, p. 11-36. 1999.

PRATES, A. P. L. *et al.* **Caring for wetlands: a response to climate change**. Brasilia: Ministry of the Environment/Secretariat of the Ramsar Convention. 2010. 27 p.

PRIMACK, R. B.; RODRIGUES, E. **Biologia da Conservaçâo**. Londrina: Editora Planta, 7ª printing. 2001.328 p.

RALEIGH, R. F.; ZUCKERMAN, L. D.; NELSON, P. C. Habitat suitability index models and instream flow suitability curves: Brown trout, revised. U.S. Fish and Wildlife Service. **Biological Report**, Washington, n. *82,* September. 1986. 65 p.

RAMSAR. **What is the Ramsar Convention on Wetlands**? **Gland: The Ramsar Convention Secretariat,** Information Paper, n. 2. 2002. 5 p.

RICHTER, B. D. *et al.* Ecologically Sustainable Water Management: managing river flows for ecological integrity. **Ecological Applications**. Tempe, v. 1, n. 13, p. 206- 224. 2003.

RICKLEFS, R. E. **The economics of nature**. Rio de Janeiro: Guanabara Koogan, 6. ed., 2010. 546p.

ROCHA, D. F. N. B. **Thermal biology of the turtles *Trachemys dorbigni* (Duméril & Bibron, 1835) and *Trachemys scripta elegans* (Wied, 1839) from the lakes of Porto Alegre (Testudines, Emydidae)**. 2005. 68 f. Dissertation (Master's Degree in Animal Biology) - Federal University of Rio Grande do Sul. Porto Alegre, 2005.

ROLOFF, G. L.; KERNOHAM, B. J. Evaluating reliability of habitat suitability index models. **Wildlife Society Bulletin**. Bethesda, v. 27, n. 4, p. 973-985. 1999.

ROSS, J. P. (Ed.). **Crocodiles**. Status Survey and Conservation Action Plan. United Kingdom: IUCN, Gland, Switzerland and Cambridge, 2. ed., IUCN/SSC Crocodile Specialist Group. 1998. 96 p.

RYDER, R. R.; MANRY, D. E. White-faced ibis (*Plegadis chihi*). In: POOLE, A.; GILL, F. (Eds.). **The birds of North America**. Ithaca and Philadelphia: Cornell Lab of Ornithology and Academy Natural Sciences. n. 130, 24 p., 1994.

SAFRAN, R. J. *et al.* Foraging site selection by nonbreeding white-faced ibis. **The Condor**. Los Angeles, n. 102, p. 211-215, 2000.

SAITO, C. H. (Coord.) **PROBIO Environmental Education**. Brasilia: Department of Ecology, University of Brasilia/MMA. 2006. 136p.

SANTOS, R. F.; CARLESSO, R. Water deficit and the morphological and physiological processes of plants. **Brazilian Journal of Agricultural and Environmental Engineering**. Campina Grande, v. 2, n. 3, p. 287-294, 1998.

SCDV. Secretariat of the Convention on Biological Diversity. **Global Biodiversity Outlook 2**. Montreal. 2006. 81 p. Available at: www.biodiv.org/GBO2. Accessed on April 2, 2013.

SCHAFER, A. E. *et al.* Geoprocessing methods as an aid in the generation of a Socio-environmental Atlas of four coastal municipalities in Rio Grande do Sul. In: XIV Simpòsio Brasileiro de Sensoriamento Remoto. 2009, Natal. **Proceedings**. p. 4377- 4384, 2009.

SCHWARZBOLD, A.; SCHAFER, A. Genesis and Morphology of the Coastal Lagoons of Rio Grande do Sul, Brazil. **Amazoniana**. Manaus, v. 1, n. 9, p. 87-104. 1984.

SETE, L. R. **Relations between the Taim Ecological Station and the Local**

**Community: Conflicts and Cooperation**. 2010. 127 f. Dissertation (Master's Degree in Rural Extension) - Federal University of Santa Maria. Santa Maria. 2010.

SHAFFER, J. A. Pattern and potential causes of White-faced Ibis, *Plegadis chihi*, establishment in the northern prairie and parkland region of North America. **Canadian Field Naturalist**. Ottawa, v.121, p. 46-57. 2007.

SICK, H. **Birds in Brazil:** a natural history. Princeton: University Press. 1993. 932 p.

SICK, H. **Ornitologia Brasileira**. Rio de Janeiro: Nova Fronteira, 3ª printing. 1997. 912 p.

SILVA, D. D.; PRUSKI, F. F. **Gestao de recursos hidricos - Aspectos legais, econômicos, administrativos e sociais**. Brasilia: Secretaria de Recursos Hidricos, Universidade Federal de Viçosa, Associaçâo Brasileira de Recursos Hidricos. 2000. 659 p.

SIQUEIRA, M. F. **Use of fundamental niche modeling to evaluate the geographic distribution pattern of plant species**. 2005. 106 f. Thesis (Doctorate in Environmental Engineering Sciences) - Sâo Carlos School of Engineering, University of Sâo Paulo: Sâo Carlos, 2005.

SKIDDS, D. E.; GOLET, F. C. Estimating hydroperiod suitability for breeding amphibians in southern Rhode Island seasonal forest ponds. **Wetlands Ecology and Management**. Dordrecht, v. 13, p. 349-366. 2005.

SOUZA, A. S. M. C. 73f. **Population genetic characterization and kinship in tapicuru, *Plegadis chihi* (Vieillot, 1817) (Threskiornithidae), from Rio Grande do Sul**. Dissertation (Master's Degree in Genetics and Evolution) - Federal University of São Carlos. Sâo Carlos, 2011. 73 f.

TARBOTON, K. C. *et al.* **Habitat suitability indices for evaluating water management alternatives**. Office of Modelling Technical Report, South Florida Water Management District. Florida, 2004. 148p.

TASSI, R. **Hydro-environmental Management of Wetlands**. 2008. 240 f. Thesis (Doctorate in Water Resources and Environmental Sanitation) - Federal University of Rio Grande do Sul, Porto Alegre, 2008.

TROST, C. H. White-faced Ibis (*Plegadis chihi*). In: (CLARK, T. W *et al.*, (Eds.). **Rare, Sensitive, and Threatened Species of the Greater Yellowstone Ecosystem**. Wyoming: Northern Rockies Conservation Cooperative/Montana Natural Heritage Program/The Nature Conservancy-Idaho.1989. 153 p.

TUCCI, C. E. M. (Org.). **Hydrological Behavior of the Taim**. Porto Alegre: IPH/UFRGS/IBAMA, v. 1 and 2. 1996.

U.S. FISH AND WILDLIFE SERVICE. **Federal Register**. Endangered and Threatened Wildlife and Plants; Reclassification of the U.S. Breeding Population of the Wood Stork From Endangered to Threatened. Department of the Interior. Final Rule. Part III. Washington, v. 79, n. 125, June 30, 2014.

U.S. FISH AND WILDLIFE SERVICE. **Habitat a Basis for Environmental Assessment**. Washington: Division Ecological Services/U.S. Fish and Wildlife Service Department of Interior, September, 1980. 29 p.

U.S. FISH AND WILDLIFE SERVICE. **South Florida Multi-Species Recovery Plan for the threatened and endangered species**. Atlanta, 1999. Available at:< http://www.fws.gov/VeroBeach/ListedSpeciesMSRP.html >. Accessed on:August 13, 2014>.

VALLE, M. P. **Sociability and territoriality in a population of *Chauna torquata* (Oken, 1816) (Aves, Anseriformes, Anhimidae) in the Taim Ecological Station**. 1991.158 f. Dissertation (Master's Degree in Zoology) - Federal University of Paranâ. Curitiba, 1991.

VAN METER, V. B. **Florida's Wood Storks**. Miami: Florida Power and Light Company, 1985. 26 p.

VERDADE, L. M.; LAVORENTI, A.; PACKER, I. U. Reproductive management of the yellow-papered caiman (*Caiman latirostris*) in captivity. In: III Workshop on the Conservation and Management of the Yellow-bellied *Caiman* (*Caiman latirostris*). 1993, Piracicaba. **Proceedings**. p. 143-152, 1993.

VILLANUEVA, A. O. N. *et al.* Preservation of the Taim wetland. In: XII Simpòsio de Recursos Hidricos. **Proceedings**. Vitória, p. 9-16. 1997.

VILLANUEVA, A. O. N. **Simulation of dynamic flood areas: composite channels and wetlands**. 1997. 164 f. Thesis (Doctorate in Resources

Hidricos e Saneamento Ambiental) - Federal University of Rio Grande do Sul. Porto Alegre, RS. 1997.

VILLWOCK J. A. *et al.* Geology of the Rio Grande do Sul coastal province. In: J. Rabassa (Ed.). **Quaternary of South America and Antartic Peninsula**. Rotterdam, v. 4, p. 79-97. 1986.

WAECHTER, J. L. Ecological aspects of restinga vegetation in Rio Grande do Sul, Brazil. Communications from the PUCRS Science and Technology Museum - **Botany Series**. Porto Alegre, RS. v. 1, p. 49-68. 1985.

WAECHTER, J. L.; JARENKOW, J. A. Composition and structure of the tree component in the peat forests of Taim, Rio Grande do Sul. **Biotemas**. Florianópolis, v. 11, n. 1, p. 45-69. 1998.

WETLANDS INTERNATIONAL. **Planting trees to eat fish: field experiences in wetlands and poverty reduction**. Wageningen: Wetlands International, 2009. 144 p.

ZACHARIAS, I.; DIMITRIOU, E. Identifying microclimatic, hydrologic and land use impacts on a protected wetland area by using statistical models and GIS techniques. **Mathematical and Computer Modelling**. Oxford, v. 51, p. 200205. 2010.

ZEDLER, J. B; KERCHER, S. Wetland Resources: Status, Trends,

Ecosystem Services, and Restorability. **Annual Review of Environment and Resources**. Palo Alto, v. 30, p. 39-74. 2005.

ZOHARY, T.; OSTROVSKY, I. Z. Ecological impacts of excessive water level fluctuations in stratified freshwater lakes. **Inland Waters**. United Kington, p. 47-59. 2011.

MIX
Papier aus verantwortungsvollen Quellen
Paper from responsible sources
FSC® C105338

Printed by Books on Demand GmbH, Norderstedt / Germany